THE PROBLEM OF KRAKATAO AS SEEN BY A BOTANIST

BY

C. A. BACKER

FORMERLY GOVERNMENT BOTANIST FOR THE FLORA OF JAVA

PUBLISHED BY THE AUTHOR

FOR SALE:

VISSER & Co.
WELTEVREDEN
JAVA

MARTINUS NIJHOFF
THE HAGUE
HOLLAND

ISBN 978-94-015-1997-7 ISBN 978-94-015-3173-3 (eBook)
DOI 10.1007/978-94-015-3173-3

PREFACE.

Since the eminent Treub in 1888 published the botanical results of his memorable trip to Krakatao in 1886 much has been written on the new vegetation of the island and gradually scientists have begun to believe that the consecutive authors have contributed to the solution of the problem how the island has covered itself anew with plants and from where these plants came.

I also once shared that belief. But now I am of another opinion for which I give the grounds hereafter. To the unprejudiced and critical reader it will be clear that I have not contented myself with demolishing old errors but have assiduously tried to find out the truth, to separate the good wheat — so scarce alas — from the huge amount of chaff. For the rest, let this paper speak for itself.

I finish by tendering my heart-felt thanks to my learned friend Dr. A. Fitz for the inestimable and disinterested assistance lent to me in translating the original Dutch text into English and to Messrs. Van Ingen at Sourabaya for their lively interest and loyal assistance in printing this book.

C. A. Backer.

CONTENTS.

Plate I

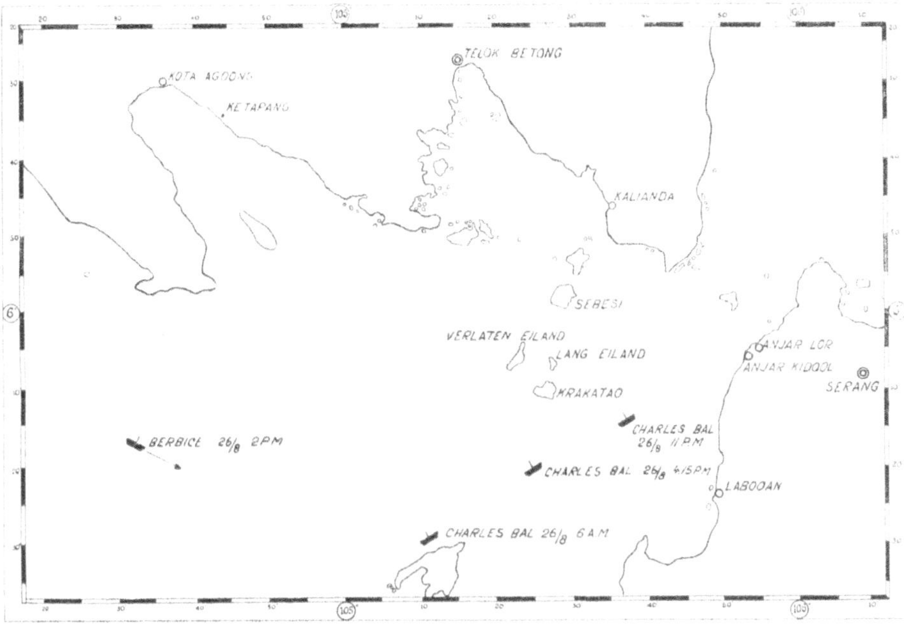

MAP OF THE SUNDA-STRAITS
SCALE 1 : 1.800 000.

Plate II

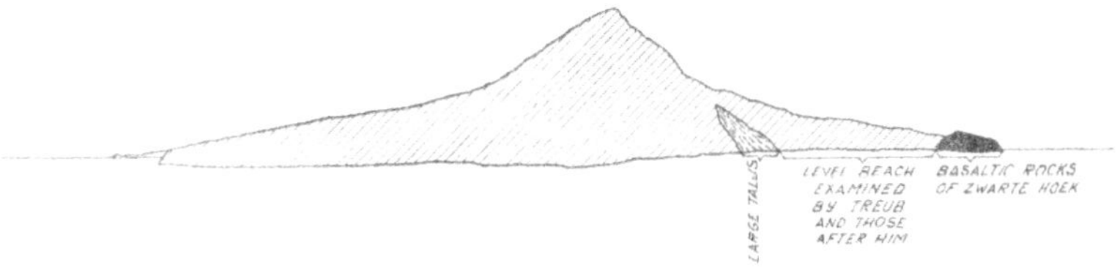

KRAKATAO SEEN FROM LANG EILAND
(ON THEIR UPPER HALF THE SLOPES ARE MUCH STEEPER THAN BENEATH)

Plate III

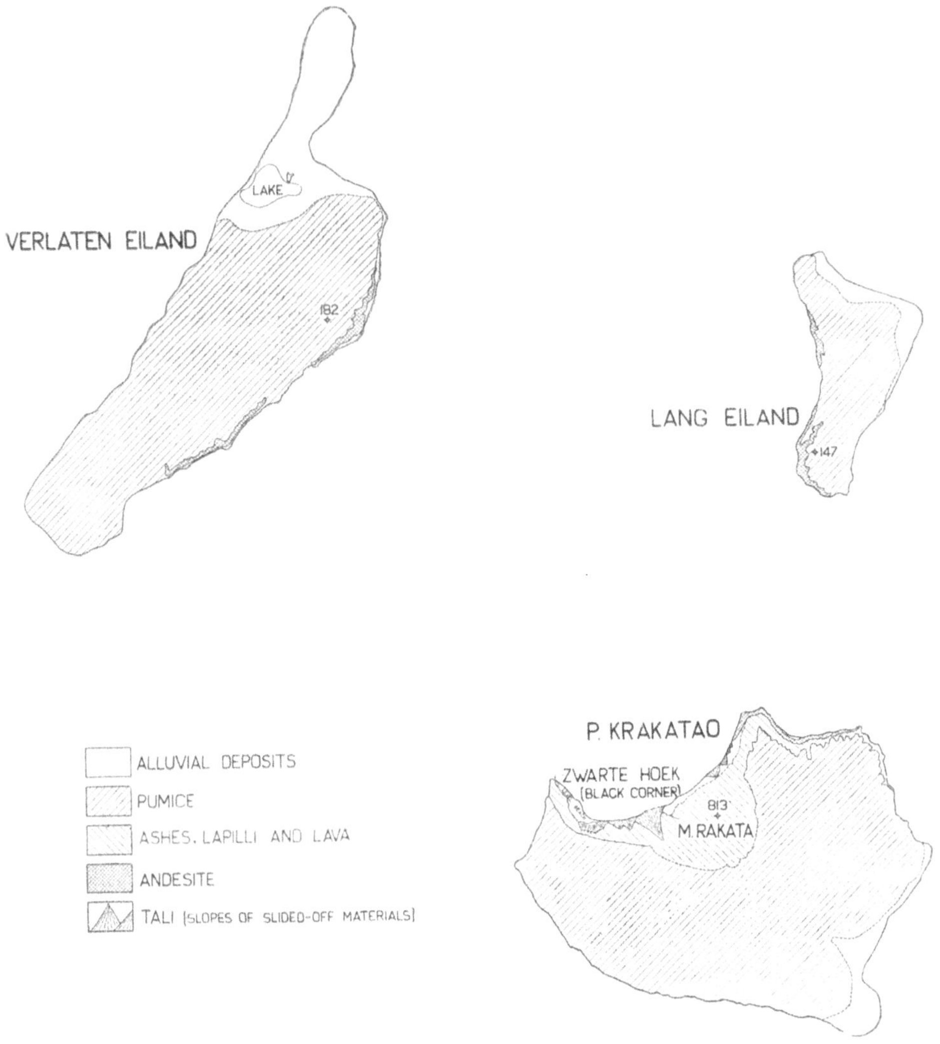

KRAKATAO-GROUP.
(IN 1908).
SCALE 1 : 100.000.

INTRODUCTION.

Whenever we have been furnished a fetish and have been taught to believe in it, and love it and worship it, and refrain from examining it, there is no evidence, howsoever clear and strong, that can persuade us to withdraw from it our loyalty and our devotion. MARK TWAIN.

Plus qu'ailleurs, le botaniste dans un pays tropical doit se munir de prudence et de méthode dans ses recherches. Mieux vaut, dans ce cas, être trop sceptique que de ne l'être pas assez. TREUB.

The volcanic island Krakatao, in August 1883 devastated by a terrific eruption, has greatly interested botanists and zoologists ever since Treub in 1888 [1]) published the results of his botanical excursion to that island in 1886. This almost universal interest was due to the fact that, on the authority of Treub, it was considered an irrefutable truth that the original flora and fauna of Krakatao had been wholly destroyed by the eruption. Hence it was thought that Krakatao presented the exeedingly rare occasion of tracing, step by step, how the flora and fauna are restored on an island, entirely deprived of its animal and vegetable life.

Treub [2]) already was perfectly aware that in this question the *cardo quaestionis* lay in the word *„how"*. Considering what may be daily observed on coral-islands it might safely be expected that, after some lapse of time, plants and animals would reappear on Krakatao. Nor in this *expectation* itself did the interest of the scientific world concentrate, neither in the *fact* of its fulfilment. But confining ourselves to the botanical side of the problem, the all-important questions were: *When* will plants reestablish themselves on Krakatao; *which species* will be the first to arrive, which will follow; *from where* will these plants come; *in which form* (seeds, spores, grown-up plants) and *by which expedient* will they cross the sea; at which point of the island will they *arrive;* which *possibilities* or *difficulties of existence* will be offered by the spot reached by them; will they, under the combined influence of soil and climate, either thrive and develop or come to a standstill and die out; which *associations will they form;* how will they, mutually, *keep up the struggle for life*, assist *each other in that struggle* or, perhaps, *gradually make their own existence impossible;* which phases of

1) Ann. Jard. Bot. Buitenzorg, 1me Série VII (1888) 213 seq.

2) Ann. Jard. Bot. Buitenzorg, 1me Série VII (1888) 213: „ . . . . . parce qu'on ne sait absolument rien sur la manière dont île volcanique se recouvre d'une végetation entièrement nouvelle".

development will accordingly the new flora, in the course of time, show? To these and other questions in the domain of floristic, oecological and genetical plantgeography botanists hoped to get an answer, *not* based on guesses, suppositions and possibilities, but on an extensive knowledge of facts, acquired by continuous and accurate research.

After the excursion to Krakatao by Treub botanists and zoologists have repeatedly visited the island, though with much longer intervals than one would expect, considering the universally acknowledged importance of the problem. Several of those visitors have published the real or supposed results of their investigations. Gradually the botanical litterature on Krakatao has grown rather voluminous but until now the botanical data have not yet been critically sifted and compiled.

Some criticism, however, does not seem out of place. Botanists sufficiently acquainted with the vegetation of the Dutch East Indies and possessing some insight into geo-botanical problems, must when studying the different publications necessarily come to the conclusion that not only the above-mentioned authors were greatly handicapped by lack of a thorough knowledge of the flora and its behaviour towards the factors of the habitat, but also that both method and intensity of the investigations never were such that they could guarantee the reliability of the results. Moreover, from these results conclusions were not always drawn with the caution required by science. Apparently some authors have been so free as to make quite hazardous guesses. As a rule, the authors have not viewed the work of their predecessors and that of their own with the keen spirit of criticism; consequently the botanical litterature has gained more in quantity than in quality by some of the papers published.

In this paper I firstly intend to examine whether the basis of discussion of all the authors working on the new flora of Krakatao vid. the hypothesis of the total destruction of the original flora of the island, is an irrefutably proven fact; in the second place to sum up the positive data – thus neglecting the more or less plausible guesses – which in the course of time have been collected about the manner in which the vegetation of Krakatao has been restored; to ascertain the real results of the successive excursions and to conclude which reliable answers they have given to the questions put forward above. But some discussions, partly of a general nature partly more special ones, have to go beforehand.

A phyto-geographical investigation in the Dutch East Indies has to overcome a score of difficulties which in the temperate zone either do not present themselves at all, or, if they do, only to a very slight extent. Complete modern handbooks enabling the non-specialist to obtain the indispensable knowledge of the flora without spending much valuable time, do not exist. This lack, the immense number of species, and the great variability which many of these show, create very serious difficulties to the phyto-geographer when naming his plants and it is clear that a correct naming of the plants is one of the fundamentals of phyto-geographical studies. The determination of sterile material offers still greater difficulties. Unhappily, sterile plants, as a rule, form the greater part of the materials the phyto-geographer in the tropics has to work with. Everyone who has made an extensive floristic investigation in the Dutch East Indian jungles, has obtained the experience that, in a given locality, only a small part of the vegetation bears flowers or fruits, a rule especially applying to the shrubs and trees which generally predominate. Often the trees are so high that materials from them cannot be obtained. The comparative scarcity of flowers also causes that most not flowering or not fruiting plants remain unobserved if one does not examine the vegetation scrupulously; amidst the luxuriant green they escape the eye. To a yet higher degree this rule holds good for persons insufficiently acquainted with tropical forms; he who does not *know* plants does not *see* them. Not small is, furthermore, the number of plants which appear above-ground only during part of the year so that in localities where, at a given time, they may abound, a few months afterwards not a single specimen is to be seen. Very patient and minute investigations, reiterated in all seasons, are required in order to draw any reliable conclusions.

Another serious difficulty, not to be underestimated, is our lack of knowledge regarding the relations between the tropical vegetation and soil, climate and other oecological factors. At the present time, now that soil and climate of Java are rather better known than formerly, this difficulty might be partly removed by studying extensive and properly labelled collections. But both manner of labelling and completeness of herbaria in the tropics leave, as a rule, much to be desired. The following complaint of Rikli [1]) is quite appropriate to many of the collections in these herbaria: „Alle früheren Forschungsreisenden bringen in ihren Sammlungen und Aufzeichnungen beinahe gar

1) Fortschritte Naturwiss. Forschung III (1911) 214.

keine Standortsangaben oder dann nur solche, die viel zu allgemein gehalten sind ...... Ja selbst bis weit in das XIX. Jahrhundert lässt die sachverständige Etikettierung der Herbarien noch vielfach sehr zu wünschen übrig. Für eine in jeder Hinsicht brauchbare Verwertung einer Belegpflanze bedarf es zum mindesten einer genauen Angabe von Standortsverhältnissen und Fundort sowie des Sammlers und des Datums ...... Bei Gebirgspflanzen ist die Höhenangabe durchaus erforderlich".

Treub and his successor to the directorate of the Buitenzorg Botanical Institute, Dr. J. C. Koningsberger, may justly claim the honour, not only of having acknowledged these very serious shortcomings and the necessity of improvements, but also of having strongly supported extensive researches on the distribution of plants and the circumstances under which they were growing. Most important though still only partly elaborated data have been collected during the latest years by two divisions of the Department of Agriculture, vid. the Forest Research Station and the Museum for Economic Botany, both in Buitenzorg. These two divisions have taken over an important part of the task of the Buitenzorg Botanical Institute ('s Lands Plantentuin) by instituting on a large scale well-conducted investigations on the forest flora and on the useful plants of the Dutch East Indies. All these researches have tended to increase and deepen our knowledge of the flora, but by making new collections the fatal consequences of former errors cannot always be eliminated.

Treub in 1888 [1]), Penzig in 1902 [2]), Ernst in 1907 [3]) already pointed out the difficulties connected with a botanical exploration of Krakatao, difficulties greatly increased by the unfavourable situation of the island in the midst of the often rather turbulent Sunda-straits, far from the ports Batavia and Tandjong Priok. Small native sailing-vessels, the only ones available throughout the year without being too costly, succeed in some seasons only with much trouble in reaching Krakatao or cannot reach it at all. This was experienced in 1921 by a Swedish botanist, Dr. Olof Arrhenius, who, starting from the west-coast of Java, made use of this kind of craft, but after three weary days had to go back without having been able to effect his purpose. Formerly the only means of reaching Krakatao with certainty were small government steam-ships, which were to be used after special permission of the authorities, but these vessels

[1]) Ann. Jard. Bot. Buitenzorg, 1e Série VII (1888), p. 215.
[2]) Ann. Jard. Bot. Buitenzorg, XVIII (1902), p. 96.
[3]) Ernst, Neue Flora Vulkaninsel Krakatau (1907), p. 33.

could only wait a few days at the island, a time quite insufficient for a close botanical investigation. For many years already the Laboratory for Marine Investigations at Batavia (ressorting under the Botanical Institute of Buitenzorg) has disposed of a steam-ship on behalf of the voyages of the officials of that laboratory, but trips to Krakatao are so very costly that, when not necessary for the service, they can but seldom be made.

Other difficulties arise from the nature of the island. This is about 14 km^2. in size and strongly accidented, with the exception of a few level beaches. Krakatao, by its large size already rather unsuitable for a minute investigation, is ploughed in all directions by innumerable deep (down to 40 m.) and very precipitous ravines which at every moment hinder or arrest the penetration of the island; the investigation is made still more difficult by the locally often very dense, almost impenetrable vegetation, by the humid heat and by hosts of most aggressive ants and mosquitoes. There is, at the surface at least, no water, there is no living man, hence neither food nor coolies are available. Without coolies an extensive botanical investigation in very accidented localities in the tropics cannot possibly be carried out. Researches satisfying moderate scientific claims would be of many weeks' duration and should be often reiterated and therefore would be very expensive.

This juncture of unfavourable circumstances is the principal cause why until now *botanical investigations were only made in few and comparatively restricted areas of the island* and often with long intervals, sometimes of 10 years and longer, and also why *these investigations always were most superficial.* An exploration satisfactory to scientific demands was never yet made.

The long intervals between the successive investigations and the fragmentary and most superficial character of these have caused that we do not know with sufficient accuracy how the vegetation of Krakatao was composed in the first decennia after the eruption, which period was the most important one for the investigations. For in the tropics also, the flora shows the phenomenon that in a not too restricted locality, that may be considered as an oecological unity, besides some dominant or constant species there always are accessory ones. Of this fact Junghuhn[1]) already gave an instance *(Exacum)* in his description of the alang-fields. The phenomenon is best to be observed in localities where a state of equilibrium has not

[1]) Junghuhn, Java, Second Dutch Ed. I, 291; German Ed. I, 213.

yet been reached. In the Indies it is a fact of common occurrence that in a locality deprived of its original vegetation, a new and at first most unstable flora rapidly develops. When a wood has been felled, when an exhausted hooma (dry ricefield) is left to its fate, when an old tea-plantation of which good crops are no more to be expected, is deserted and gradually turns into jungle, when ravines are filled up by lahars (streams of sand or mud), when a plot of ground is raised by spouting on it mud or sand from the bottom of the sea, when new land is reclaimed from the waters, the areas deserted by culture or newly gained quickly cover themselves with plants. In this case there are always a few species [1]) especially grasses, ferns and *Compositae* which will predominate within a short time, the sooner as they are more common in the neighbourhood and dispose of better means of transport. But amidst the dominant species there always appear some herbs, shrubs and trees which, at first only accessory, become gradually constant or dominant, whilst the species formerly dominant gradually disappear. He who will make in a suchlike locality a study of the development of the new, yet unstable flora has to reckon with this fact. Therefore an extensive investigation of the whole of the locality is required; it is not enough to examine at intervals some small spot where at the time of investigation not a single specimen may occur of species, momentarily accessory, but which have been dominant or constant or will become so in future.

Just the same case presented itself at Krakatao. There also, from the very first, a number of accessory species were scattered among the dominant ones. T r e u b already drew attention to the scarcity of Phanerogams in the lower parts of the island as contrasting with the great number of Cryptogams. In the older litterature on Krakatao we come across a striking proof of the presence in the island of only locally occurring plants. T r e u b collected in 1886 on the northwestern side of the island a very conspicuous fern, *Onychium siliculosum* C. Chr. (= O. *auratum* Kaulf.). In March 1897 a small company of botanists, among whom were T r e u b, the well-known pteridologist R a c i b o r s k i and P e n z i g, visited the same side of the island. Of course they tried to find back the plants of T r e u b. On this occasion P e n z i g and R a c i b o r s k i followed two different

[1]) In Java, especially: *Ageratum conyzoides L.* and *mexicanum Sims,-Baltimora recta L.* – *Bidens pilosus L.* var. *albus Schulz,* – *Clibadium surinamense L.* – *Erechthites hieracifolia Raf.* and *valerianifolia Raf.,* – *Eupatorium pallescens D.C.* – *Gleichenia laevigata Hook.* and *linearis Clarke,* – *Imperata cylindrica Cyr.,* – *Lantana Camara L.* – *Piper aduncum L.* – *Saccharum spontaneum L.* and *Trema orientale Bl.* – On the sea-side often *Chloris barbata Sw., Pluchea indica Less.* and *Paspalum vaginatum Swartz.*

ridges, which, however, were almost parallel and so close to each other, that a person shouting on the one ridge was heard and understood on the other. Penzig who was accompanied by Treub (who had found the plant 11 years before) expressly states that *Onychium* was *not* found back [1]), Raciborski on the contrary declares of the same trip and the same plant: *Very common in the island Krakatao* [2]). Doubtless both were, in some way, right; on the one ridge the fern may have grown copiously, on the other it will have been absent or very scarce. Quite analogous phenomena may be observed everywhere in the Javanese mountains and hills and often also in the plains [3]).

Since the vegetation of two adjacent ridges of Krakatao differed at the same time so much that a species absent or very rare on the one ridge was copious on the other, much caution is required in comparing the results of consecutive excursions. If one is not absolutely sure of having investigated the same locality under the same climatological conditions as one's predecessor and with sufficient accuracy, one cannot possibly know whether in the time elapsed between two investigations a given species has disappeared or made its first appearance, whether it has become more common than formerly or otherwise. The two first explorations of Krakatao exclusively took place on the northwestern side, consisting for a large part of basaltic rocks, partly bare; the three subsequent investigations were made not only there but also on the *lower* eastern and southeastern slopes of the island, which almost everywhere are covered by a thick layer of ashes and pumice. Considering that two adjacent ridges differed already in vegetation, it is extremely probable that the south and south-eastern slopes, lying wide apart form the north-western side and differing moreover geologically, have shown important differences in the flora. But until now, *authors on the flora of Krakatao have simply lumped together the results obtained in different localities* and at different

---

1) Ann. Jard. Bot. Buitenzorg XVIII (1902), p. 108.

2) Raciborski. Die Pteridophyten der Flora von Buitenzorg (1898), p. 164. A specimen collected in Krakatao by Raciborski is present in the Buitenzorg Herbarium and there is another one collected in March 1905 by Valeton in the same island.

3) A remarkable instance of this fact was observed by Dr. L. den Berger and the author in April 1920 south of the village Bantoor in Eastern Java. From this village two roads run in a southern direction at a small distance from each other. Along the eastern road *Priva lappulacea Pers.*, a herb introduced form tropical America, was growing in innumerable specimens; along the other one, apparently presenting the same conditions, we only found a single specimen after having sought a long time. — In some inundated ricefields of Java *Xyris indica L.* is so very numerous that the whole field is tinged with yellow whilst in adjacent fields of apparently the same condition no specimen is to be seen. The same experience is often made with *Hydrolea zeylanica L.* and *Ammannia octandra Roxb.* Many more instances might be added.

times. When a certain species was not recorded for a trip it was concluded that it did not occur in the island at the time that trip was made. Repeatedly, also when the investigations ware *not* made in the same locality, one may read in the publications [1]) that plants were not present at the time of a former expedition, that they have recently appeared, that they have become more common or less frequent, that the character of the vegetation was modified since the last investigation.

It need scarcely to be said that a mere list of *names* of species, which were found at a few trips made with large intervals, is of little value for an investigation on the gradual renewal of a quite or almost destroyed flora. If such a list shall be of value for the study of the vegetation, it has to be not only practically complete but moreover it must be accompanied by, or made as a part of a thorough oecological and phyto-sociological investigation.

As appears not only from notes added to the publications on Krakatao but also from the very character of these publications itself, the new flora of the island has never yet been investigated by persons who had made a sufficient study of the flora of the Dutch East Indies and were able to recognize *on the spot* all or nearly all plants, and could consequently make out the frequency of each of the species. *All* investigators contented themselves with collecting some specimens or fragments of the species they chanced to observe *of which they knew at best the names and as a rule not even so much;* most, if not all, of the materials were named after the trip was finished, as a rule at Buitenzorg. To this I would raise no insurmountable objection if only in the collections made in this manner every species was represented by a sufficient number of forms, if subsequent collectors had always seriously studied all the materials gathered by their predecessors in order to make themselves, as far as possible, acquainted with the flora and, finally, if they had occasion of revisiting Krakatao with short intervals in order to resume and extend their investigations after having worked out the results of a former trip. These favourable circumstances, however, have never yet coincided: during or after each new excursion it appeared that the collectors did not know or did not recognize many of the plants which had been found already at former trips and often in large numbers. Moreover, not a single publication shows that

---

([1] Ernst, Neue Flora Vulkaninsel Krakatau (1907), pp. 8, 46; — Docters van Leeuwen in Ann. Jard. Bot. Buitenzorg XXXI (1921), pp. 58, 117.

in the field the necessary attention was paid to oecological factors. It was never carefully examined and noted in what kind of habitat each species lived and how it behaved there, neither in which numbers it occurred. The superterranean and subterranean development of the plants, the degree of weathering, the texture, structure, watercapacity and microflora of the soil, the temperature, the intensity of light, the mutual relations between flora and fauna, were never yet seriously investigated. Many of the scientific sides of the problem were insufficiently attended to or entirely neglected and the consequences of this neglect did not fail to appear.

It is only fair to admit that in 1886, when the eminent savant Treub, who was no florist and whose knowledge of tropical plants was very limited, made his ever-memorable trip, the Buitenzorg Herbarium was in a sad state. The collections were very incomplete, they were badly named and worse labelled; hence by studying them nobody could obtain a sufficient knowledge of the Indian vegetation and the factors determining it. Information dating from that time about habitat or frequency of the different species is for the greater part unreliable. The comparatively little material, then extant, had as a rule, not been collected in the manner it should be done, it was often in a bad state of preservation and wrongly named; data about altitude above sea-level, character of the habitat and local frequency were often entirely lacking. In later years the situation has grown much better. An investigator provided with scientific insight, with a sense for criticism, with sufficient knowledge and experience, and not sparing labours can, at the present time, find in the Buitenzorg collections many reliable data about habitat and local frequency of a multitude of species, which data will allow him to draw conclusions in the domain of oecology. Therefore one is quite justified to claim from the recent publications on the new flora of Krakatao a more *exhaustive* and especially a more *critical* elaboration of the subject than might be reasonably expected from the former ones.

Doubtless a very important part of the Krakatao-problem lies in the answer to the question by which means the present flora, which all authors agree to be new [1]), has reached the island. It is

1) When this paper was about to go to the press the following reference in Botanisches Centralblatt, Neue Folge 11 (Band 153) 1927, p. 155 came to my notice: Scharff, R.F., Sur le problème de l' île de Krakatau. C. R. Congr. Ass. franç. Grenoble (Biogéogr.) 1925 4 S. Nach kurzer Schilderung der Eruptionen und ihrer Wirkung äuszert Verf. seine Zweifel daran, dasz Pflanzen- und Tierleben 1883 vollständig vernichtet worden sind, also die heutige Lebewelt der Insel insgesamt durch passive Verbreitung nach dort gelangt sei. Für eine ganze Reihe von subterranen und Felsspalten bewohnende Formen weist er die Möglichkeit des Überlebens einer

most remarkable that on this point (the littoral flora excepted) never any actual observation or experiment was made; all authors confined themselves to more or less happy guesswork, thereby guided by the old prescription: nearly all Cryptogams and the Phanerogams possessing light or winged seeds were carried over by winds, the plants with fleshy or succulent fruits by birds or bats, the rest by the sea. The possibility of human influence was but seldom thought of. Treub [1]) went so far as to strongly deny this possibility. But in the tropics as well as everywhere else may be observed that the influence of man on the wild flora is by no means a slight one. This fact may be easily accounted for: how often do not light fruits, provided with hairs, hooks or viscid organs, adhere to garments! How easily seeds and fruits may be conveyed along with luggage! After the eruption, but previous to the excursion of Treub in 1886 Krakatao had already been repeatedly visited by man [2]); in 1883 by Verbeek in the company of 4 Europeans, in August 1884 by Verbeek with 2 Europeans, in September 1884 by Verbeek with 11 Europeans, in July 1885 by Verbeek with 6 Europeans. All these persons with their native servants and their belongings were embarked at Batavia or Tandjong Priok, in the neighbourhood of which ports the flora is very rich. It could hardly be avoided, that at these excursions seeds or fruits were carried along. The natives who are not in the habit of scrupulously cleansing their articles of dress go ashore both in Java and on Krakatao; luggage is placed on the ground at the places of embarkment and disembarkment: to garments, to footgear, to luggage, to articles of food seeds may easily have adhered.

It is not at all sure, not even probable, that landings on Krakatao were exclusively made by Europeans and their native followers. In the Java-sea fishers in order to cook or eat their meals or to dry the fishes caught are wont to go ashore at uninhabited islands, where driftwood is always plentiful. Before the eruption of 1883 Krakatao was now and then visited by Sumatran fishers [3]) and there

solchen Katastrophe nach; die „neu auftretenden" Pflanzen sind nicht, wie bei Verschleppung zu erwarten gewesen wäre, zuerst am Ufer, sondern in der Inselmitte aufgetreten und somit erscheint das ganze Problem der Neubesiedelung von Krakatau, Verlaten und Lang Eiland in seiner notwendigen Voraussetzung, der Tatsache der Neubesiedelung in Frage gestellt – I have not been able to consult the short paper in question but am glad to state that its author quite independent of me has come to the same conclusions as myself.

1) Ann. Jard. Bot. Buitenzorg, 1me Série VII (1888), p. 215.

2) Of this I was informed by Dr. Verbeek, (letter of April 25th 1922).

3) Verbeek, Kort Verslag, p. 6. Van den Berg in Tijdschrift voor Indische Taal-, Land- en Volkenkunde XXIX (1884), p. 223. – Dr. Arrhenius kindly told me that in 1921 the Wedana (native district-chief) of Anjer Kidool informed him, that fishermen were then in the habit of landing on Krakatao.

is no reason why they should not have done so afterwards. By these people also plants may have been introduced, either accidentally or purposely. That Cocos *nucifera* was introduced in this manner is out of doubt (Cf. Chapter IV.).

One mostly is inclined to exaggerate the share which the wind takes in the transport of *seeds* and *fruits* across the sea. Anyhow, not for a single one of the species occurring at present on Krakatao it has been *proven* that its spores, seeds or fruits were carried over by wind, though for the minute spores of Cryptogams and seeds of orchids I will by no means deny the *possibility* of this conveyance. One should bear in mind that, by land, seeds and fruits are much more easily carried over long distances by wind than across the sea. On land, where light objects after having dropped, may be swept up again by every strong gust of wind and where the wind has often a more or less upward direction, due to obstacles in its way, strong and continuous air-currents can probably convey very light seeds and fruits over long distances. But this does not hold good for transport across the sea. Seeds and fruits fallen into the water can no more be carried upwards but either remain floating on the surface or sink to the bottom. In the former case further conveyance can only take place by sea. In the latter case they are lost unless they happen to fall into the water quite near the coast, where they may be washed ashore. Transport of seeds and fruits of some weight by wind across a water-level can therefore practically be made only in a single flight: that this has happened in the case of Krakatao, vid. over a distance of more than 20 km., has never yet been proven. The reality of this transport, however, has always been considered an axiom. Rightly Diels[1]) observes: „Dasz der Phantasie ein groszer Spielraum bleibt, auch beträchtlichere Entfernungen auf diese Weise überbrückt zu sehen, bedarf keiner Erwähnung und bestätigt sich in der Literatur". The presence of ferns and *Compositae* on Krakatao is by no means conclusive evidence for transport by wind as I will show in my discussion on the paper of Treub.

Transport of light seeds, which different authors on the flora of Krakatao have ascribed to the action of winds, might in many, if not all, cases have been as well made in another manner. Numerous fruits, seeds and spores either fall directly into watercourses or are carried along to them by rain-water flowing rapidly over the ground during the heavy showers which are so frequent in West-Java and

[1]) Prof. Dr. L. Diels, Pflanzengeographie (1918), p. 43.

South-Sumatra. By these watercourses seeds are, for a part at least, conveyed to the sea which may carry them yet further. It has never been proven that after having reached the sea all these seeds soon sink to the bottom or lose their germinative power [1]). In the tidal creeks of West-Java the water is often locality tinged brown by millions of floating spores of the typical mangrove-fern *Acrostichum aureum* L. When the tide is low these spores are carried off to the sea and it is not probable that all of them soon perish there. It is a well-known fact that light seeds readily adhere to driftwood, to pumice or to other floating objects. In this way spores and seeds might have been carried over. In the Westmonsoon when the rivers discharging their water in the Sunda-straits are often much swollen by continuous heavy rains, they not seldom carry eradicated trees far into the sea. Seeds, fruits or epiphytes adhering to the emerging parts of these trees can in this way reach Krakatao *within the time of a single day* [2]). Ernst [3]) found in 1906 two fungi growing on a trunk washed ashore; he supposes that the mycelium, which had formed new reproductive bodies, had survived the transport, but this opinion, however plausible, is not corroborated by proofs.

Schimper already expressed the opinion that in the tropics winds take but a small part in the dispersal of seeds of littoral plants. The arguments alleged by Ernst [4]) against this opinion are far from conclusive. The assertion of Ernst: „Von ausserordentlicher Bedeutung für die Enstehung der neuen Krakatao-flora ist, wie schon durch den ersten Besuch Treub's in überraschender Weise gezeigt worden ist. . . . . der Wind", is unsufficiently accounted for. *The suggestions* Treub *made on this point of the problem prove nothing at all.* Rikli [5]) gives a striking instance how easily one may be mistaken when considering winds as a means of conveyance over long distances: „Ein kleiner Vorfall aus dem Jahre 1903 zeigt, wie bei der Beurteilung der Verbreitung von Blättern durch den Wind grösste Vorsicht geboten ist. Bei der Ueberschreitung der vorderen Furka zwischen dem Pommat (Italiën) und dem Val Bosco (Tessin) fanden wir in einer Höhe von 2300 m. Buchenblätter. Da die obersten

[1]) Birger (Beihefte zum Botanischen Centralblatt XXI (1907), p. 263, seq.) has shown that many seeds of European non-halophytes after having been submerged during several days in a salt solution of 3,4% had retained their germinative power. I do not know of any experiment made on this point with the seeds of the plants found on Krakatao.

[2]) Ernst, Neue Flora Vulkaninsel Krakatau (1907), p. 56.

[3]) Ernst, Neue Flora Vulkaninsel Krakatau (1907), p. 58.

[4]) Ernst, Neue Flora Vulkaninsel Krakatau (1907), p. 62.

[5]) Rikli, in Fortschritte Naturwiss. Forschung III (1911), p. 259.

Buchen im Gebiet sich in der Nähe der Ausmündung des Val Campo befinden, schlossen wir auf einen Windtransport von reichlich 9 km. und über eine Höhendifferenz von 1200 m. Am folgenden Tag begegnete uns aber beim Abstieg ein Mann der einen Sack voll Buchenlaub trug und der auf unsere Frage wozu und wohin zur Antwort gab: „Zum drufliege auf d'Alp". Damit schrumpfte der vermeintliche grosse Windtransport auf kaum 1,5 km. zusammen und wie leicht dürfte nicht an den rauhen Kleidern der Sennen das Laub weiter verschleppt werden, so dass schiesslich für den eigentlichen Windtransport recht wenig übrig bleibt".

Concerning import by animals, it is a well-proven fact that fruits and seeds often adhere to feathers or feet of birds and, in this manner, may be carried over long distances. It is most probable that the avifauna of Krakatao has been entirely driven away or destroyed by the very violent eruption of 1883, which was of some month's duration. The many birds found in the island after the eruption may have spread seeds endozoically or exozoically. In the first months after the eruption a zealous research after seeds found on birds or in their droppings could have given a reliable indication that plants were *introduced* in this manner. But such a research was then never made. At the present time, now, that almost the whole of the island is covered with a dense vegetation and the flora is but very imperfectly known, one cannot possibly make out whether such seeds originate from the island itself or from elsewhere. Hence every assertion that plants were introduced by birds is a mere *supposition,* nothing more. The possibility may exist but proofs are entirely wanting. In my discussion on the different papers on Krakatao I shall revert to this subject. At present I only once more may emphasize the fact that *not one of the investigators has made any conclusive experiment* in order to ascertain whether spores, seeds or fruits were carried over from elsewhere to Krakatao either by winds or by animals.

Summarized, my general objections against the investigations and publications concerning the renewal of the flora of Krakatao are the following: By situation, size and nature the island is unsuitable for an extensive and accurate investigation; up to now only very small parts have been investigated and the investigations were always most superficial; the intervals between the successive investigations were much too long; the investigators did not possess the required knowledge of the East-Indian flora and the oecological factors; after the origin and manner of introduction of the new flora many guesses have been made but no serious investigation has taken place; no attention was paid in the field to oecological factors and many other important points; the investigations were often made in different localities, so that an accurate comparison of the successive results becomes impossible. Doubtless very valid excuses may be advanced for the incompleteness of the investigations. *But it is inexcusable that on such grounds conclusions were based which could only be justified if they had been founded on scrupulous researches.*

II. THE VEGETATION OF KRAKATAO BEFORE THE CATASTROPHE OF AUGUST 27th 1883.

Of the vegetation of Krakatao before the catastrophe but very little is known. From a series of old communications, collected by Mr. N. P. van den Berg [1]), I derive the following data: Wouter Schouten mentions the high, densely wooded island of Krakatao which he passed in October 1685. In February 1681 J. W. Vogel saw to his astonishment „dass die Insul Cracketow „so bei meiner Hinreise gantz grün und lustig mit Bäumen sich „präsentirte nunmehr als gantz verbrannt und wüst vor unsern Augen „lag, und an vier Orthen grosze Feuerblocke aus wurffe. Und als „ich den Schiff-Capitain befragte, zur welcher Zeit ermeldete Insel „gesprungen, so berichtete er mich, das solches in May des 1680 Jahres geschehen".

It is, however, worthy of note that none of the log-books of the numerous ships that in May 1680 and the previous and following months passed the Sunda-straits, records an eruption of Krakatao. In May 1681, only 3 months after the great devastation was noticed by Vogel, Elias Hesse observed [2]) „that Krakatao and all the other islands in the midst of the Sunda-straits are covered with big trees and jungles". Hence probably Van den Berg rightly supposes that Vogel has greatly exaggerated and that the eruption of 1680 can have been only a small one. Presently I shall mention the lava-streams in the then still extant northern part of the island and it is perhaps not quite impossible that these and the basaltic rocks of Zwarte Hoek (*Black Corner*) were then formed. In February 1780 the home-bound ships of the explorer James Cook, who had been murdered on Hawaii a year before (Febr. 14th. 1779), called on Krakatao. On the southside the sailors found a small brook and moreover a thermal spring where the few inhabitants of the islands came to bathe. The island was covered with woods, barring some spots where the natives had cleared the forest and planted rice.

For yet a long time afterwards Krakatao was inhabited. In 1809 Horsfield speaks of a village near a brook where the ships

---

1) Mr. N. P. van den Berg, Vroegere berichten omtrent Krakatau, in Tijdschr. voor Indische Taal-, Land- en Volkenkunde XXIX (1884) p. 208 seq.

2) „dat Cracatouw en de andere eilanden, welke allen midden in Straat van Sunda leggen, voorsien syn met hoge bomen en wildernissen.

were provided with water, firewood, goats, poultry and fruits. Afterwards Krakatao served as a place of banishment for criminals from the Lampongs (South-Sumatra) and when this was stopped the island remained entirely deserted and was only from time to time visited by fishermen.

So far the data collected by Van den Berg. Botanical data just as vague as the preceding ones are given by Junghuhn [1]) who says no more than: „From its base to its very crown Mount *Rakata*" (the highest, at the same time the southern-most and, at present, the only top of the island) „is covered by woods". A confirmation and partly a correction of this statement is given by Verbeek [2]) who visited Krakatao in July 1880, about 3 years before the eruption. He found the southern part of the island coverd with such a dense vegetation, that it would have been very difficult to penetrate into the interior. But in the northern part [3]) he saw several rather steep lava-streams which bore only a very meagre vegetation and which apparently belonged to the youngest and least weathered rock-formations of the island.

Teysmann, the famous chief-gardener of the Buitenzorg Botanical Gardens must have visited Krakatao some decennia before the eruption; the exact year is unknown to me and is of no importance, but cannot lie far from 1860. Five plants, collected by him on the island, are recorded in Miquel, Flora Indiae Batavae, Sumatra (1862). These are:

Loranthaceae

1. *Loranthus pentandrus L* (= Dendrophthoë pentandra Bl).
2. *Viscum articulatum Burm.*

Leguminosae

3. *Intsia amboinensis Thouars.*
4. *Mucuna gigantea D. C.*

Meliaceae

5. *Dysoxylum arborescens Miq.* (= Hartighsea acuminata Miq).

That is all. In truth, meagre reapings, quite insufficient to give an idea of the composition of the old vegetation. Of the 5 species named, No. 1 and 2 are parasites, very common in the plains of

1) Junghuhn, Java 2nd Dutch Edition II, 4; German Edition II, 3.
2) Verbeek, Krakatau, Dutch Edition (1888), p. 7; French Edition (1885), p. 7.
3) Verbeek, Krakatau, Dutch Edition (1888), p. 5, p. 164; French Edition (1885), p. 5.

Java, the second, as a rule, hyperparasitizing on the first [1]. No. 3 and 4 are inhabitants of the beach and of low, level coral-islands. No. 5 is a tree occurring in the Javanese forests from the plains up to an altitude of 1400 m. It is therefore quite possible that Teysmann collected only on the shore and immediately behind it; the list of his plants gives no indication as to whether he penetrated into the interior of the island. Consequently nothing at all is known of the former flora of the interior and especially of the top. It would not be difficult to do some guess-work, but about Krakatao already far too many guesses were ventured forth.

In the northern part of Krakatao a series of eruptions began on Sunday, May 20th 1883 which about 3 months afterwards (Monday, Aug. 27th 1883) culminated in and ended by the blowing up and plunging into the sea of the entire island barring the southern half of the Rakata-cone. In this cone there was also a crater, *but this took no part in the eruption.* On August 10th and 11th, when on the northern part of Krakatao only a few naked tree-trunks [2] were left standing, *the entire southern half of the cone still bore a dense vegetation* [3]. Two weeks afterwards this cone was covered by ashes and pumice which, — as is almost generally believed up to the present time, — entirely destroyed the vegetation [4].

---

1) On *Rutaceae* (Euodia) Viscum articulatum is sometimes found as a primary parasite.

2) Verbeek, Krakatau, Dutch Edition (1888) p. 26; French Edition (1925) p. 27.

3) Verbeek, Krakatau, Dutch Edition (1888) p. 28; French Edition (1925), p 28.

4) See however page 9, footnote 1.

III. THE DESTRUCTION OF THE ORIGINAL VEGETATION OF KRAKATAO.

Has it been proven that in the still extant part of *Krakatao* the vegetation was entirely destroyed by the eruption of 1883? Consequently that nowhere living plants, rhizomes, seeds, spores, mycelia or bacteria were left on or in the soil. Has the soil really been sterilized?

The answer to this question determines the direction of the researches on the origin and the development of the new flora. Is this flora entirely new and should its origin exclusively be sought outside of Krakatao or do we also find there the remnants, or their posterity, of the vegetation that formerly covered the island? And if so, which part of the present vegetation has to be considered new, which might be old?

Treub was fully aware, that one of the essential points of the problem lies in the question whether the vegetation was *totally* or only *partly* destroyed. In his opinion the total destruction was fully proven; he even considered it impossible that the results of further researches ever could outweigh the strong reasons whereupon he meant to have based his opinion. His argumentation runs as follows [1]:

„Les données que je puis fournir sur la nouvelle flora de Krakatau sont dûment acquises dès aujourd'hui; il n'y a pas lieu de craindre qu'elles seront infirmées par des recherches ultérieures".

. . . .

„En premier lieu il est indispensable de prouver que la flore actuelle doit être considérée comme nouvelle et qu'elle n'est pas issue des restes de la végétation luxuriante qui occupait l'île avant l'éruption. Rien n'est plus facile d'ailleurs, que de fournir cette preuve. Lors de l'éruption les arbres terrassés ou brisés par de violents rafales ont dû être à moitié carbonisés, vu la température extrêmement élevée qui a certainement régné sur toute l'île. Ensuite Krakatau a été couvert, depuis le sommet jusque bien au delà du niveau de la mer, d'une couche de cendre et de pierre ponce brûlantes. Cette couche a une épaisseur variant entre un mètre et soixante mètres. Dans ces conditions là il est clair qu'aucun vestige de la flore n'a pu subsister après le catastrophe. La graine la plus persistante et le rhizome le mieux protégé ont dû perdre à jamais toute vitalité.

[1]) Ann. Jard. Bot. Buitenzorg, 1me Série VII (1888), 213, 214, 215.

Point n'est besoin de discuter la possibilité que les plantes nouvellement installées à Krakatau y soient amenées par l'intermédiaire de l'homme. L'île est inhabitée, inhabitable et difficile à visiter.

Enfin la composition elle-même de la flore actuelle de Krakatau, démontre une fois de plus qu'elle n'a pu tirer son origine ni de la flore antérieure ni par l'intervention de l'homme".

In order to rightly value the conclusive force of this reasoning we have to examine its bases. Treub advanced two arguments for his belief in the total destruction of the flora. The first holds that the destruction was the inevitable consequence of the island being covered by a dense and thick layer of glowing eruptive products; the second deduces the destruction of the old vegetation from the composition of the new one in 1886 Both of these arguments will here be examined.

Has it indeed irrefutably been proven that in 1883 the entire surface of Krakatao was covered with a layer of ashes and pumice [1], thick and hot enough to destroy all vegetable life? This has so universally been taken for granted during more than forty years that the time has come to consider whether the belief is in harmony with the facts. In the first place we have to examine whether Treub carried out his investigations on a scale, large enough that his conclusions may be considered to be valid for the entire island. I refer here to what I have said in the first chapter (p. 5) about size and nature of Krakatao.

It is strange that Treub himself did not mention which part of the island was investigated by him. In order to get information on this point I have applied to the renowned Krakatao-specialist Verbeek (whom I quoted already before as I will do many times more) in whose company Treub made his excursion. From letters of Verbeek it appears [2] that Treub did not reach the higher parts of the island but restricted his investigations to the lower localities on the northwestern side near Zwarte Hoek. Verbeek assures that at this time no other plants were growing there than these mentioned by Treub [3].

[1]) Many authors, i.a. Schimper (Pflanzengeographie (1908) p. 90), Penzig (Ann. Jard. Bot. Buitenzorg XVIII (1902) p. 103) and Warming (Pflanzengeographie, 3te. Aufl. (1918) p. 688) wrongly think that in 1883 Krakatao was covered by lava also. But during the eruption of that year no lava has flown out. (Cf. also Ernst. Neue Flora Vulkan-insel Krakatau (1907) p. 4). The present lava-streams are part of the old mantle of the volcano.

[2]) Letters of April 25th 1922 and May 7th 1923.

[3]) In Ann. Jard. Bot. Buitenzorg, 1me Série VII (1888) p. 213 seq.

Verbeek, whose extraordinary merits in the domain of geology I respectfully acknowledge, may of course not be considered an authority in the domain of botany but yet I think he may be right here. Treub's description of the situation seems to me to be mainly correct, but only for so far as concerns part of the localities investigated. The lower parts of the island were really almost everywhere covered with a very hot layer of pumice many meters thick, from the deeply eroded ravines [1]) of which, two months after the eruption, hot water and steam emerged in several places. Verbeek who visited Krakatao 5 times with short intervals (Oct. '83; Aug. '84; Sept. '84; July '85; June '86) told me by letter: „Two months after „the eruption the masses of pumice were still very hot; everywhere „steam escaped from small crevices; the barefooted natives tripped „when coming near or on such a crevice". Therefore it is highly probable that in those lower parts of Krakatao which remained deeply buried, the original vegetation was destroyed, partly by suffocation, partly by the high temperature to which it was exposed during a long time. As late as 1908 my companions and myself [2]), when visiting the lower localities on the southwestern side of the island, got the impression that the vegetation, years before, had been entirely destroyed.

But we may ask: Did not Treub and the subsequent authors generalize too much? Is it not possible that the higher areas, not reached by Treub, differed from those beneath?

In my opinion both questions have to be answered in the affirmative. I have here in view especially the localities on the south- and south-eastern slopes of the cone, upwards of $\pm$ 500 m. above sea-level. By their situation (see map) they were protected against the violence of the eruption; two weeks before the end of the eruption they still bore a dense vegetation (Cf. p. 17) and at the final catastrophe they were not covered with a thick layer of coarse blocks of pumice as the areas beneath, but only with a much thinner layer of ashes and fine grit [3]). Near the top the layer of eruptive products was even *very thin* [4]). Two months after the

1) Verbeek, Kort Verslag 15,— Krakatau, Dutch Edition (1888) 302.

2) Jaarverslag van den Topographischen Dienst in Nederlandsch Indië (Year-book of the Topographical Survey Service in the Dutch Indies) 1908, p. 189.

3) Verbeek, Krakatau, Dutch Edition (1888), p. 122.

4) Verbeek, Krakatau, Dutch Edition (1888), p. 376. How firmly the fiction of a total and very thick covering has rooted in the brains of scientists appears i.a. from a communication of Dammerman [Treubia III (1923) p. 61] who, in spite of the above-mentioned very clear statements by Verbeek, covers the entire island with a layer of hot ashes of 30—60 m. thickness.

eruption the original soil was found to lie locally bare in the eroded ravines [1]).

A thin covering, lasting no more (and probably rather less) than 2 months, may have destroyed many superterranean organisms, but must be considered insufficient to kill all deeply buried rhizomes, roots, seeds, spores, mycelia and bacteria, unless it can be proven that it has strongly heated the underground to a considerable depth and during a long time.

But for the *thinly covered* higher areas on the south- and south-eastern sides this proof has not been given. During the first years after the eruption the temperature of the underground was never measured [2]). In 1883 no lava has flown out and the locally thin layer of ashes and fine grit need not at all to have been very hot. The products ejected by the crater, active in 1883, — I recall to memory that the crater in the Rakata-cone itself remained inactive —, possessed during the catastrophe an initial speed [3]) of more than 1000 m. per second. As the active crater lay far beneath the top of the Rakata-cone on its northern side and at a distance of but a few km., only such eruptive products as were almost vertically ejected could fall down on the averse side of the cone; the other ones must have been hurled far beyond it. The materials which were thrown up almost vertically must have reached a very considerable height, partly of 50 km. and more [4]), and thereby very cold zones of the atmosphere. The mean air-temperature [5]) in the Dutch Indies at 5 km. above sea-level is $-1,5°$ C, at 10 km. $-34°$ C, at 15 km. $-74°$ C, at 20 km. $-80°$ C. Upwards of 20 km. the temperature is somewhat higher but remains still very low. At 30 km. it is $-55°$ C and at 50 km. probably as much. Above 30 km. there are strong air-currents which, during the eruption, must have had a speed of 30 m. per second. This speed is, at this altitude, in the latest years also observed by means of pilot-balloons.

The ashes and the fine grit which reached a greater height

1) Verbeek, Kort Verslag (1884), p. 13.-Krakatau (1888) p. 122. In 1908 (Jaarverslag Topogr. Dienst Ned.-Indië 1908, p. 157) Fransen Herderschee found at a low altitude a ravine with a rocky bottom, which made the impression to be an *old* lavastream and afterwards, beneath an altitude of 300 m. another ravine the bottom of which was formed by the original rock-formation, the old mantle of the volcano. In 1916 De Jongh (Jaarboek van het Mijnwezen in Nederlandsch Oost-Indië 1916, p. 43) stated that a not unimportant part of Krakatau was deprived of its tuff-mantle and exhibited the original surface of the cone.

2) Verbeek, by letter dated April 25th 1922.

3) Verbeek, Krakatau, Dutch Edition (1888), p. 123.

4) Verbeek, Krakatau, Dutch Edition (1888), p. 123.

5) Communicated to me by letter of Dr Boerema, subdirector of the Meteorological Observatory of Weltevreden (Batavia).

than 30 km. were probably carried by these strong air-currents far from their place of origin. But the fine material that has fallen on Krakatao and was probably not thrown up higher than 20 à 30 km. has reached the coldest air-zones, from −80° C tot −55° C. In vacuo this very light material would have needed somewhat more than a minute to come down, but due to the resistance of the air this time must in reality have been very much longer. Its falling speed and hence also the heat caused by friction must have been low. This fine material, though very hot when ejected, was during a long time surrounded by extremely cold air and had plenty of time to cool. This is the very plausible explanation given by Verbeek [1]) of the phenomenon that, whilst the ashes, fallen during the eruption near Ketimbang in Sumatra at a very great distance from Krakatao, ejected under a small angle of elevation and having reached neither a great height nor very cold air, came down so hot that they caused severe burns and even set fire to native houses [2]), the ashes and small pieces of pumice fallen near Krakatao were by far not so hot and did not cause burns anywhere [3]).

At a comparatively small distance (15–20 km.) from Krakatao the falling ashes and grit rendered the air warm and stifling [4]) but they caused nowhere serious trouble. The pieces of pumice which fell on a ship (Charles Bal) less than 20 km. distant from Krakatao [5]), and which „covered the decks to three or four inches very speedily" even those „of a considerable size and quite warm" were so harmless that the sailors could sufficiently protect themselves with a southwester [6]); therefore they cannot have had a really high temperature. But another situation presented itself on a ship (Berbice) at a distance between 70 and 90 km. from Krakatao, where the larger pieces of pumice burned holes in garments and sails [7]).

1) Verbeek, Krakatau, Dutch Edition (1888) p. 124, 125.

2) Under the pseudonym „Miles", Freiherr Von und Zu Egloffstein, head-manager of the tea-plantation Tjikopo near Buitenzorg wrote [Deutsche Wacht, VIII (1922), No. 1 p. 30] about the little village Minanga in South- Sumatra, 80 km. distant from Krakatao, which village he visited a few days after the eruption: „Die meisten Häuser waren verbrannt: auch waren viele der Bewohner ein Opfer der glühenden Asche geworden. . . . Schon von weitem entdeckte ich eine ganze Reihe Tragbahren. Es waren dies die mit schweren Brandwunden bedeckten Schlachtopfer des glühenden Aschenregens".

3) The fact that ashes ejected by a crater came down in the neighbourbood much cooled was i.a. observed in 1817, during the violent eruptjon of Mount Idjen (East-Java). Cf. Junghuhn, Java, 2nd Dutch Edition II, 710: „Die Asche war. . . . dort und überall als *trockner und kalter* Aschenregen gefallen.

4) Verbeek, Krakatau, Dutch Edition (1888), p. 324.

5) See plate I.

6) Verbeek, Krakatau, Dutch Edition (1888), pp. 69 and 95.

7) Verbeek, Krakatau, Dutch Edition (1888), pp. 71 and 97.

The closer to the crater the finer materials came down, the lower was their temperature [1]).

Together with the ashes were ejected innumerable large blocks [2]) of pumice, containing very hot gases. As with increase of the size of objects, the surface and hence also the resistance of the air and the eradiation of heat increase in ratio much less than the weight, these blocks remained a much shorter time in the cold air-zones and lost – by a double cause – much less heat: they were probably somewhat cooled on the outside, but must have been still very hot within when come down [3]). Such large heavy blocks which from a height of thirty or perhaps fifty km. fell down at an enormous speed could not come to a standstill on the higher, steeper slopes [4]) of the cone, but, as is generally observed [5]) during volcanic eruptions, leapt and rolled down until they reached the lower less steep parts where they came to rest. Consequently these lower parts have been much more deeply buried than the higher ones. One can distinguish on the south and south-eastern sides of Krakatao two zones, of course not clearly demarcated: the higher parts above 400–600 m. which were covered only with a rather thin and not very hot layer of ashes and grit [6]), and the lower, more gently sloping parts covered

---

1) Of course I do not mean such very fine eruptive products as were shot up to a very great height, were there cooled very much, and carried away by strong winds, fell down far from Krakatao.

2) Verbeek [Krakatau, Dutch Edition (1888), p. 122] saw on Krakatao blocks of 1 cubic metre in size. De Iongh (Jaarboek van het Mijnwezen in Nederlandsch Oost-Indië 1908 No. 42) roughly estimates that the eruptive products consisted for 15–25 % of pieces of pumice varying in size between that of a pea and of a human head, for 3–5 % of blocks of hard rock, for the rest of volcanic ashes.

3) The initial temperature of the ejected materials must have been more than 800° C. It seems [Cf. Kemmerling, Vulkanol. Meded. II (1921), p. 78] that volcanic gases are enclosed in the pumice and can escape only very slowly.

4) See plate II made after plate XXVII in Jaarverslag Topogr. Dienst Ned. Indië, 1908, and cf. Ernst, Neue Flora Vulkaninsel Krakatau (1907) tab. III, fig. 4. Cf. also Ernst, l. c. p. 28: „Nach Südosten senkt sich der halbierte Kegelberg steil gegen einen flacheren Fuss."-De Iongh (Jaarboek van het Mijnwezen in Nederlandsch Oost-Indië 1916, p. 45) expresses the opinion that during the eruption the ashes fallen on the higher part of the cone, slided off.

5) Cf. i. a. the description by Junghuhn, Java, 2nd Dutch Edition III, 783, 1120; German Edition II, 542, 760) of the eruptions of Mount Smeroe and Mount Lamongan. Also Kemmerling, Vulkanol. Meded. No. 3, p. 4

6) The statement of Docters van Leeuwen [Ann. Jard. Bot. Buitenzorg XXXII (1923) 139] and Dammerman (Treubia II, 73) that the ashes fallen on Krakatao must have been much hotter than these fallen on Sebesi about 20 km. to the north, is consequently incompatible with well-stated facts. The great heat on Krakatau was not caused by ashes but by hot blocks of pumice. On Sebesi no very big blocks of pumice-stone have fallen. According to Verbeek [Krakatao, Dutch Edition (1888), p. 122] on the latter island pieces as big as a human head are already scarce. „Generally speaking the large and heavy blocks of pumice fell within a circle of 15 km. radius around Krakatau. The materials fallen outside this circle consist for the greater part of ashes". Therefore the vegetation of Sebesi suffered much less than that of most parts of Krakatao.

not only with fine materials but also with a very thick layer of coarse pumice, which preserved its warmth during a long time. Both zones are clearly marked on the map added to the Transactions of the Congress at Weltevreden (1919), of which plate III is a reduced copy.

The *superterranean* vegetation was probably almost or entirely destroyed over the whole surface of the island, except in very well sheltered localities. Verbeek, who two months after the eruption ascended the mountain on its western side, close to the brink of the rupture [1]), did not see a single erect tree, not even a dead one, but only fallen and carbonized trunks [2]). *But it is proven that not everywhere the wood was burned.* From 1915 till 1917 (Cf. Chapter X) Krakatao was inhabited by Mr. Händl who had obtained a concession for winning volcanic products on the island. Mr. Händl sank many shafts in order to measure the thickness of the layer of pumice, beneath which he found in many places *unburnt* treetrunks and other parts of plants. Undoubtedly great devastations were brought about by the big, hot blocks of pumice, which fell everywhere at enormous speed in great numbers, rolled and leapt down the sides of the cone and must have smashed everything in their way. The forest may – locally – have taken fire. Leaves and thin twigs may have been broken off by the weight of the ashes and the grit. Violent squalls, such as repeatedly occur during eruptions and which during the catastrophe were observed in the Sunda-straits [3]) and South-Sumatra [4]) may, as Treub [5]) already observed, have hurled trees against the ground. During the eruption of Mount Gĕdè in November and December 1840 large parts of the forest were destroyed in this manner [6]). The same happened in 1846 on Mount Tangkuban Prahu [7]), in 1901 and 1919 on Mount Kelut [8]). The pumice-rain may have caused scorching vapor-blasts such as are mentioned for Mount Kelut by Junghuhn [9]) and are said to have come forth from that mountain in 1919 also [10]). Prob-

---

1) Verbeek, Kort Verslag, 13,— Krakatau, Dutch Edition (1888) p. 475.

2) This observation proves that, locally, the covering layer had not been very thick. Verbeek apparently did not examine *to which depth* the trunks had been carbonized.

3) Verbeek, Krakatau, Dutch Edition (1888), 68, 468, 469; French Edition (1885) 71.

4) Verbeek, Krakatau, Dutch Edition (1888), 55; French Edition (1885), 57.

5) Ann. Jard. Bot. Buitenzorg, Ime Série VII, 214.

6) Junghuhn, Java, 2nd Dutch Edition II, 40,— German Edition II, 31.

7) Junghuhn, Java, 2nd Dutch Edition II, 60.— German Edition II, 46.

8) Vorderman in Teysmannia XII (1903), p. 276. Kemmerling, Vulkanologische Mededeelingen No. 2 (1921). p. 41.

9) Junghuhn, Java, 2nd Dutch Edition III, 708, 709.— German Edition II, 492,

10) Kemmerling, Vulkanologische Mededeelingen No. II, (1921), 42,

ably several of these factors have co-operated to destroy the superterranean vegetation almost completely.

But all this proves by no means that everywhere the soil was heated so much and to such a depth that all *sub*terranean parts of the vegetation were killed. In the higher regions the underground *was protected against being heated too much by the comparatively cold ashes and fine grit which continually fell down*. The large blocks, glowing hot within, but already somewhat cooled on the outside, did not remain above but rolled and leapt rapidly down till they reached the lower parts. There only they can have strongly heated the soil [1]); on the higher parts of the cone they did not remain long enough to do so. For pumice and ashes are very slow conductors of heat, and heat does not soon penetrate deep into the soil as may be observed everywhere. After every forest-fire it appears that the flames, though having destroyed many or all superterranean organs and having more or less deeply carbonized the trunks of the trees, have not entirely killed the vegetation. A number of plants always survive by their subterranean organs and, some time after the fire, everywhere young growth appears from roots, rhizomes or seeds, that have remained in life. On May 23rd 1901 Mount Kelut ejected glowing hot vapor-blasts and ashes that killed several people and locally scorched the superterranean vegetation. V o r d e r m a n, when visiting a week afterwards the spot where the calamity had occurred, observed [2]) that a scorched specimen of *Amomum* then already was forming new leaves, which shows that the heat cannot have penetrated deep into the soil.

During several hours on the last day of the eruption (Aug. 27th 1883) a drizzling rain of cold humid ashes fell in West-Java and South-Sumatra [3]) and caused a considerable reduction of the temperature [4]). On the interjacent island Krakatao most probably the same has happened. If so, this cold rain will have contributed to reduce the temperature of the fallen materials. Wherever the covering layer was thin, the temperature of it must have fallen perceptibly.

Taking all this into consideration I cannot admit that from the

---

1) A few days after the eruption of Mount Kelut in 1919 K e m m e r l i n g (Vulkanol. Meded. II (1921), pp. 77, 78) observed in and near the crater in a loose mass of pumice, washed together by a flood of water, temperatures of ± 200° C and a fortnight afterwards yet of 165° C.

2) Teysmannia XII (1902), p. 276.

3) V e r b e e k, Krakatau. Dutch Edition (1888), pp. 45, 51, 96, — French Edition (1885) pp. 47, 53, 100.

4) V e r b e e k, Krakatau, Dutch Edition II (1885), 324.

history of the eruption itself may be concluded that the vegetation of the island *must* have been entirely destroyed by heat or by suffocation. On the contrary, everywhere in the higher regions where the covering layer was but thin and soon washed away, where on steep ravine-sides far projecting rocks sheltered the vegetation beneath, moreover in deep hollows or crevices of precipitous cliffs, lower soil-organisms, mycelia, spores, seeds, rhizomes, root-systems and even entire plants, when not too large, may have survived. They will undoubtedly have suffered from the drought, caused by the soil having been somewhat heated, but after sufficient rain had fallen, many remnants may have begun a new life.

Consequently it is not in the least proven that by the eruption of 1886 the soil of Krakatao was everywhere thoroughly sterilized. From the very first the researches were conducted by Treub in a predefined and, in my opinion, wrong direction. As he positively meant that the entire island had been covered with a permanent thick layer of very hot eruptive products, he felt quite convinced that all life had been destroyed and therefore considered it superfluous to assiduously seek for surviving rests of the old vegetation. Had he known that his argument was based on a fallacy, the sharp-witted savant would have acted otherwise.

If, indeed, rests of the old flora have survived in the higher ravines on the south- and south-eastern sides, we may presume that these ravines have become starting-points of the new vegetation. The eruption took place towards the close of August, in the very midst of the dry monsoon. On Krakatao the east-monsoon is, as a rule, interrupted by repeated rains [1]) as appears not only from the ravines having been deeply eroded already two months after the eruption, but also from the present vegetation [2]). Yet it is probable that

---

1) Of Krakatao itself no definite figures about the amount of rain exist. In the 3 nearest rainfall-stations, Anjer Kidool, Kalianda and Telok Betong, which all lie in the plains (see plate I), the 4 driest months of the year [3]) have together resp. 25.3, 33 and 38.5 rain-days. Hence in the *dry monsoon*, on an average, a period of 2–4 dry days alternates with a rain-day. The total amounts in these 4 months are resp. 322, 428 and 484 mm., so that the mean fall per rain-day ($12^1/_2$ à 13 mm.) is by no means insignificant. The supposition of Ernst [Neue Flora Vulkaninsel Krakatau (1907), p. 50] that on Krakatao, which rises up to above 800 m. and therefore most probably receives more rain than the plains around, in the east-monsoon, rains are comparatively scarce, is not probable, much less proven.

2) This is testified by many facts: As late as in 1908 on the porous pumice-ridges xerophytes did not or only most sparingly occur; *Saccharum spontaneum* was growing gregariously on the ridges; *Spathoglottis plicata* and *Arundina speciosa* were copious on steep ravine-sides; *Cyrtandra sulcata* was numerous at a low altitude; many fungi and epiphytic ferns and Orchidaceae have been found in the island.

3) Cf. Boerema, Rainfall in the Netherlands Indies (1925), pp. 2 and 164.

some months had to pass before the declivities with their very permeable soil had received and stored a sufficient quantity of water to allow a somewhat quick revival of the sorely damaged vegetation. Then, however, in the deeply eroded higher ravines young forest will gradually have developed, issuing from the former woods but poorer than these, all non-resistant species having been killed. This forest, at first very thin, gradually became denser. As soon as its components began to fructify the forest extended wherever the circumstances were favorable to its growth, soonest perhaps in a downward direction, as many fruits must have been carried forth by rainwater flowing down along the ravines. The ridges being drier and sunnier than the ravines will have been clothed only much later with forest, if such has happened at all. I think that in such a manner *part* of the new vegetation has developed. For that, on the seashore at least, it developed also in another way, is beyond doubt.

Verbeek, who, from his ship, a year after the eruption, saw – or meant to see [1]) – on Krakatao some grass-leaves did not search the above-mentioned ravines. In the first year the young vegetation of trees and shrubs could not have grown to such a size that a non-botanist could observe it from afar. But three years afterwards Treub [2]) could, from his vessel, already distinguish plants near the top of the cone (which reaches up to more than 800 m.) from which fact one may safely conclude that these plants were neither very small nor very young. Penzig and his companions in 1897 examined only the lower parts of the north-western side of the mountain, which they reached from the north; hence they could not possibly know anything of the vegetation in the higher ravines on the south- and south-eastern sides. Only as late as 23 years after the eruption, the first as yet very meagre data [3]) about

---

1) Verbeek told me by letter of May 7th 1923: „The grass-leaves which, when navigating along the southern shore in August or September 1884, I meant to see, judging from the green colour of some tufts, caused astonishment to me as everything else was gray and bare. *But on that occasion I did not go ashore there.*"

2) Ann. Jard. Bot., 1me Série VII (1888), p. 214.

3) Cf. Ernst, Neue Vulkaninsel Krakatau (1907), p. 28. The following communication of Ernst Haeckel [4]) is too vague to be of value for the study of the development of the new flora: „When passing in the afternoon of January 23th 1901 near the southern coast of Krakatao we could distinctly see not only the narrow brown lavaridges (sic!) which radiated from the top of the cone in every direction and pierced its green mantle, but also, on the shore, a small grove of trees which seemed to reach a height of 10–15 m". Anyhow, from this communication appears that already in 1901 the higher ravines were densely clothed with vegetation.

4) Malayische Reisebriefe. The original not being at my disposal I have had to translate from the Dutch translation (Juynboll, Uit Insulinde, p. 212).

the vegetation of these ravines were obtained and these data strongly prove for the correctness of my views. When Ernst, Pulle, Campbell and myself in April 1906 approached Krakatao from its southern side we distinctly saw how the Barringtonia- and Casuarina-associations formed a narrow strip along the shore. Behind this strip extended a broad grass-jungle with scattered trees. But in the upper half of the mountain several ravines were clothed with woods running up high against the cone; the upmost ridges and the top bore scattered trees. These forests in the ravines are considered by me to be rests of the old vegetation or the off-spring of these.

As one may read in Ernst's publication [1]) we did not succeed then in reaching these higher ravines. In the short time at our disposal and with our very defective equipment we could not penetrate the broad, very dense, 3–5 m. high grass-jungle [2]), despite of the utmost exertions.

But during the expedition in 1908 my companions and I reached a height of ± 400 m. and at the same time the ravine-forests, at the height of 200–400 m. still rather young and, so to say, forming the vanguard of the older, further developed woods in the higher parts [3]). In the very few ravines I had occasion to search, this young forest mainly consisted of *Ficus fistulosa Reinw., Ficus fulva Reinw., Ficus toxicaria L.* and *Homalanthus populnea O. K.* with an undergrowth of *Leucosyke capitellata Wedd., Cyrtandra sulcata Bl., Heckeria umbellata Kth.* var. *subpeltata. D. C., Leea indica Merr.* and *Selaginella plana Hieron.* All of these plants are frequent in forests at low altitudes in West-Java; it is quite possible that they were growing on Krakatao before 1883 and survived the eruption. Only a single day being available in 1908 for the botanical investigation of this side of the mountain, it necessarily was very fragmentary and superficial. Referring to what I have said on pp. 5 and 6 about the composition of unstable vegetations in the tropics, I think it quite possible, even very probable, that in other ravines than those searched by us, or in the higher parts of these, the young forest consisted entirely or partly of other plants. That in 1919 several species were collected on Krakatao that had not been found in 1908, is by no

[1]) Neue Flora Vulkaninsel Krakatau (1907), p. 33.

[2]) This jungle consisted mainly of *Saccharum spontaneum* L. – Junghuhn (See Chapter VI) gives a vivid description of such a *Saccharum*-jungle.

[3]) Fransen Herderschee [Jaarverslag Topogr. Dienst Ned. Indië. (Year-book of the Topographical Survey Service in the Dutch Indies) 1908, pp. 156,157] found in the lower ravines trees of 10–15 m. height. In the higher ravines the trees were taller (up to 20 m). Hence it is probable that the vegetation has proceeded from above downwards.

means proof that these species have established themselves on the island after 1908; all of them may have grown there long before, The remarks [1]) „that in 1919 in the lower ravines a great many more plants occurred than formerly", that „most of these plants were not yet present on the island at the time of the first expeditions" and „that the number of forest-plants has been greatly added to", lack a firm scientific base. It is not certain, not even probable, that in 1919 the same ravines were examined as in 1908. Of course I will not at all *deny the possibility* of new plants having immigrated after 1908, but *this is a more supposition and the Krakatao-problem did not derive its importance from the occasion to make more or less plausible guesses but from the supposed possibility to determine with certainty the origin and development of the new flora.*

I fully acknowledge that the situation in 1906 and 1908, resp. 23 and 25 years after the eruption, gives only an *indication* but no *proof* for the correctness of my views. But the total destruction of the old flora is just as little proven and, moreover, not at all probable. It is much to be regretted that Treub, considering the total destruction of the old flora as a dogma, has not seen the necessity of sending a well-equipped expedition to Krakatao, as soon as possible after the eruption. Such an expedition ought to have been repeated several times and to have been charged with instituting throughout the island and especially in the higher ravines minute investigations after rests of the old flora, which might have survived the catastrophe. This has been neglected, and therefore Treub's argumentation, based on theoretical considerations and very incomplete investigations, moreover contradicted by probability, lacks all conclusive force.

Yet another part of the present island Krakatao has not been covered with a permanent thick layer of coarse pumice-materials. I have here in view the basaltic slope, which begins at Zwarte Hoek and runs up rather high against the mountain. This locality was described by Penzig [2]) as follows: „Die z. Th. aus hartem Lavafels „bestehenden, z. Th. von Bimsstein und Lavatrümmern bedeckten „Abhänge sind sehr steil. Das Haupthinderniss" (to the ascent) „bilden „unzählige, oft auszerordentlich tiefe Spalten und Klüfte. Diese Klüfte, „wahrscheinlich durch die Contraction der erkaltenden Lava gebildet, „gähnen an ihrem obern Rande oft viele Meter weit, während die steil „abstürzenden Wände sich erst in unabsehbar dunkler Tiefe vereinigen".

[1]) Docters van Leeuwen in Handelingen Eerste Ned. Ind. Natuurwet. Congres (1920) p. 47.—Ann. du Jard. Bot. XXXI (1921), p. 58, p 117.

[2]) Ann. Jard. Bot. Buitenzorg, 2me Série, Vol. III (1902), p. 96.

These rocks were not formed by the eruption of 1883 but are part of the old mantle of Krakatao [1]). Locally the slopes are here so steep that the coarse blocks of pumice-stone, which may have fallen down upon them, could not remain where they fell, but rolled or leapt downwards, whilst the ashes and fine pumice-grit were soon washed away or slided down into the ravines or into the sea. At this corner of Krakatao the sea was filled up by the eruptive products and therefore in the first years after the eruption, till far from the coast, rather shallow [2]). Already in October 1883, two months after eruption, the old rocks of Zwarte Hoek were found to be locally bare, as appears from plate 25 in the album accompanying the work on Krakatao by Verbeek. Therefore it is quite possible, even probable, that here also, in well-sheltered localities, part of the old vegetation has survived.

Definite data about the composition of the old vegetation of these rocks do not exist, as no botanist has visited them before the eruption of 1883. But probably this locality will have borne the same flora as the old lava-streams which Verbeek saw in the no-more existing northern part of Krakatao. These lava-streams, which may have been formed by the eruption of 1680 [3]) and possibly much earlier, were after a lapse of at least two centuries yet unweathered, and bore a very scanty vegetation. A such-like, little weathered, sterile lava-stream is found at present on the eastern declivities of Mount Guntur, above Tjipanas (near Garoot), and its vegetation answers to the general description given by Verbeek. When in Java such sterile rocky localities are found in regions where the east-monsoon, as a rule, is interrupted by repeated rains and where, consequently, no extreme droughts entirely prevent the growth of plants, the rocks bear, especially in their crevices, a peculiar vegetation, which, for the rest, is almost restricted to vertical walls of hollow roads and terraces and steep riversides. Such plants,

---

[1]) Cf. Escher in Handel. Eerste Ned.-Ind. Natuurwetensch. Congres (1920), p. 205; „It appears from the topographical map of 1908 (Jaarverslag Topogr. Dienst 1908) that immediately west of Zwarte Hoek a distinct valley runs in a south-eastern direction. I found this valley to be the boundary between the old basaltic island and the young pumice-tuffs. East of the valley basalt is found down to sea-level, west of the valley pumice-tuffs down to the level of the sea". In fig. 3 of Escher's paper this is distinctly marked. A very fine plate is found in Jaarverslag Topogr. Dienst Ned.-Indië (Year-book of the Topographical Survey Service of the Dutch East-Indies) 1920, plate XXIX.

[2]) See the paper of Escher in Handel. Eerste Ned.-Ind. Natuurwetensch. Congres (1920), p. 204 and the large map accompanying it.

[3]) Verbeek, Krakatau (1888), p. 5, p. 164. If in 1680 no eruption has taken place these lava-streams must have flown out before that year.

growing by preference on steep walls and, apparently, better fit than other ones to grow in little weathered, hard, sterile localities, lacking in humus I shall indicate in future by the name of kremnophytes [1]). To the Javanese kremnophytes of the lower, not too dry regions, belong, besides a number of algae, mosses and lichens, especially ferns, i.a. *Adiantum caudatum L.*, – *Adiantum lunulatum Burm.*, – *Antrophyum latifolium Bl.*, – *Antrophyum reticulatum Kaulf.*, – *Aspidium melanocaulon Bl.*, – *Aspidium vastum Bl.*, – *Blechnum orientale L.*, – *Ceropteris calomelanos Und.* – *Cheilanthes farinosa Kaulf.*, – *Cheilanthes tenuifolia Sw.*, – *Doryopteris concolor Kuhn.*, – *Dryopteris arida O.K.*, – *Dryopteris cucullata Christ.*, – *Dryopteris intermedia O.K.*, – *Dryopteris mollis Hieron.*, – *Dryopteris setigera O.K.*, – *Gleichenia laevigata Hook.*, – *Gleichenia linearis Clarke.*, – *Hemigramma latifolium Copel.*, – *Lindsaya cultrata Sw.*, – *Nephrolepis exaltata Schott.*, – *Onychium siliculosum C. Chr.*, – *Odontosoria chinensis J. Sm.*, – *Polystichum diaphanum Moore*, – *Pteris ensiformis Burm.*, – *Pteris vittata L.*, – *Schizoloma ensifolium J. Sm.*, – *Trichomanes javanicum Bl.* and several more [2]). In sunny or slightly shadowed localities *Lycopodium cernuum L.* and the grass *Pogonatherum paniceum Hack* are very common, the latter often clothing vertical walls over large surfaces. *Oldenlandia hispida Bth.*, – *Peperomia pellucida Kth.*, – *Pilea microphylla Liebm.* and, locally, *Epithema saxatile Bl.*, – *Rhyncoglossum obliquum Bl.*, – *Monophyllaea Horsfieldii R. Br.*, – *Begonia hirtella Link* and *Salomonia cantoniensis Lour* are of rather common occurrence on steep walls. *Canscora diffusa R. Br.* and *Sonerila begoniifolia Bl.* are much rarer. Among the Orchidaceae *Arundina speciosa Bl.* and *Spathoglottis plicata Bl.* belong to the common kremnophytes. Before the eruption many of these species may have grown on steep ravine-walls and unfertile ridges on Krakatao and some individuals may have survived the eruption, either by their rhizomes or by spores or seeds, which began a new life as soon as the ashes that covered them were washed away.

It is remarkable but, from our point of view, easily accounted for, that Treub who made his researches on the above-mentioned slopes near Zwarte Hoek found there a rich fern-vegetation, composed, in the main, of part of the species recorded above. Of the 11 species of ferns collected by Treub, 2 are typical for swampy or at least

1) Derived from κρημνός, steep wall, and φυτόν, plant. Cf. Backer en Van Slooten, Javaansche Theeonkruiden, (1924) p. 14.

2) These species are by no means oecologically equivalent, especially not as regards their need of light and humidity. The species of *Antrophyum*, *Aspidium* and *Trichomanes* are heliophobic, those of *Gleichenia* heliophilous. Consequently they are never found together in the same locality.

very humid localities. These were probably [1]) collected in deeply eroded ravines, where the water-table was very high. Of the 9 other ones (see in chapter IV the corrected list of Treub's finds) no less than 7 belong to the just-mentioned kremnophytes.

This, anew, points to the possibility that the rich fern-flora found by Treub on Krakatao within 3 years after the eruption — and wich may have been present much earlier — *was not introduced from without but originated from the island itself.* I admit that this is not proven and that it never can be so, but the contrary is just as little proven. That the rich fern-flora, found by Treub, originated from surviving specimens of such ferns as may have grown for years on the rocks, is much more probable than that in the period of not yet fully 3 years between the visit of Verbeek and that of Treub, spores were carried over by the wind from far across the sea in such numbers that a rich and entirely new fern-vegetation could arise. It is quite possible that within a few years the kremnophytic vegetation was restored from its rests, that afterwards it extended from there on localities as yet covered with ashes but, on the rocks, remained as it was. Hence, Penzig in 1897 found „zum grössten Teile noch unverändert dasselbe Vegetationsbild welches Treub auf seinem ersten Besuche vor mehr als zehn Jahre antraf [2])." During the expeditions of 1906, 1908 and 1919 [3]) the base of the basaltic rocks of Zwarte Hoek still bore a very scanty vegetation; nothing is known of the vegetation of the higher parts.

Treub himself, without being aware of it, *gave a very clear hint to the possibility of the survival of part of the old flora.* He says that during his visit to Krakatao the other two islands of the group *did not bear any vegetation.* Though he did not visit, much less search these islands, and consequently had no right of speaking so decidedly, it is sure, that, if they bore a new vegetation, this was much less developed than that of Krakatao, otherwise Treub would have seen it from his small ship that could — and we may safely suppose did — sufficiently approach the shore.

The fact that on the other two islands of the Krakatao-group in 1886 a new vegetation was absent or, at least, was so little developed as to be invisible from a slight distance, can be hardly

1) Treub does not give any data about the habitat and freqency of the individual species.

2) Ann. Jard. Bot. Buitenzorg XVIII, p. 100. This communication probably originated from Treub who accompanied Penzig.

3) Docters van Leeuwen in Handelingen Eerste Ned. Ind. Natuurw. Congres (1920), p. 42.

accounted for if we are to believe that all of these 3 islands had to receive a new vegetation from across the sea. Why should the introduction of plants on the lower parts of Krakatao have taken place so much earlier than on the other two islands, which lie quite near, rise to a height of upwards of 100 m. and receive, as appears from their present vegetation, a sufficient amount of rain during the east-monsoon. *But if we admit the possibility that on Krakatao part of the old vegetation has survived the eruption, it becomes quite conceivable that on this island the flora was developed much further.* The results of the investigation of Lang Eiland by Boerlage in 1896 point into the same direction. Whilst Treub 3 years after the eruption found 11 ferns on the lower parts of Krakatao, Boerlage, 13 years after the eruption, found on Lang Eiland which he ascended to the top (147 m.) but 4 fern-species [1]). This difference seems to me too great to be quite accidental.

From the above it is clear that I decidedly refuse to accept Treub's argumentation that it appears from the composition of the flora of Krakatao in 1886 that it cannot have originated from the former vegetation. On the contrary, there are several indications that the new flora of Krakatao partly originated or, at least, *may* have originated from rests of the former one.

For the conclusive force of Treub's argumentation that man cannot have played any part in the introduction of the new flora I refer to p. 10 where I already have discussed this point. I think it quite possible, even probable, that man *did* play a part.

Against the argumentation advanced by Treub (cited by me on pp. 18, 19) I oppose the following conclusions:

1. It is entirely unproven and even unprobable that the old vegetation of Krakatao was totally destroyed by the eruption of 1883.

2. Hence a considerable part of the present flora may be the offspring of rests of the old one.

3. Man may quite well have contributed to the origin of part of the present flora.

The first and second of these conclusions are perhaps not applicable to the other two islands of the Krakatao-group, Lang

1) Viz. *Ceropteris calomelanos Und.*, — *Dryopteris unita Christ.*, — *Nephrolepis exaltata Schott*, — *Pteris vittata L.*

Eiland and Verlaten Eiland. The first of these lies north of Krakatao at a distance of somewhat less than 3 km., the other one north-west of the main island and about 6 km. distant from Zwarte Hoek. Both are smaller and much lower than Krakatao; the top of Lang Eiland reaches 147 m. above sea-level whilst Verlaten Eiland rises up to 182 m. Both islands, not as the remaining part of Krakatao partly sheltered by a mountain-cone, were fully exposed to the fury of the eruption; most probably both were entirely covered with a very thick (up to 50 m.) layer of ashes and pumice, and, to my knowledge, there are no indications that the original soil soon has been laid bare; afterwards this has happened [1]) locally. Therefore I consider it quite possible, though not irrefutably proven, that in contradistinction to Krakatao the old flora of Lang Eiland and Verlaten Eiland was entirely destroyed by suffocation and heat. I mentioned already that Treub in 1886, from his ship, observed rather numerous plants on Krakatao but on the two other islands saw no vestige of a vegetation. The first botanists who visited Lang Eiland were Boerlage and Burck who in September 1896 made a trip to this island then already for some time inhabited [2]) by some officials of the Topographical Survey Service and their native underlings; Boerlage found there a rather rich vegetation. Of the situation on Lang Eiland in 1896 Mulder gives the description cited in chapter V. — A fine picture of part of the same island 10 years afterwards is printed in Jaarverslag van den Topographischen Dienst in Nederlandsch-Indië (Year-book of the Topographical Survey Service in the Dutch Indies) 1908.

If one is right in assuming that in 1883 the original vegetation of Lang Eiland and Verlaten Eiland was entirely destroyed, these islands might have given a much more solid base (but no more than that) for investigations after the restoration of a vegetation destroyed by an eruption, than Krakatao where part of the old flora may have survived the catastrophe. Both islands have, like Krakatao, the inconvenience of embarrassing the botanical researches by the presence of innumerable very deep precipitous ravines; over Krakatao they have the advantage of being smaller (resp. $\pm$ 3 km^2. and $11^1/_2$ km^2.) and by far not so high. When studying the development of the new flora of these two islands, allowances must be made for the very

---

[1]) Cf. Hugo Cool, Krakatau in 1908, pp. 10, 12.

[2]) Not Verlaten Eiland as Ernst wrongly says [Neue Flora Vulkaninsel Krakatau (1907), p. 38] See also Tijdschr. Kon. Ned. Aardrijksk. Genootschap, XIV (1897) p. 119.

small distance which separates them from Krakatao with its probably partly saved vegetation and furthermore for the fact that Lang Eiland during several months (August 1896 till January 1897 and still later) was inhabited by persons who twice a month were in communication with Java and received from there their food and their drinking-water so that man must have influenced the flora. Finally one should bear in mind that the first investigation of the new flora of Lang Eiland took place in 1896, of that of Verlaten Eiland in 1897, 13 and 14 years after the eruption, so that we know nothing about the first and most important years. Afterwards also the researches on the flora of these islands were much more superficial and fragmentary than those carried out on Krakatao; hence the results are yet more scanty and more unreliable than those obtained on the main island and possess, if possible, yet less scientific value. Therefore when discussing the consecutive excursions, I shall separate the data obtained on Krakatao from those on the other islands, I shall restrict myself to studying the results obtained on the main island and only incidentally refer to the other two islands which, if proper investigations had been instituted, perhaps should have formed the main object. Now it is much too late, we know nothing with certainty about the origin and development of their new flora, and we never shall know.

IV. THE EXPEDITION OF TREUB IN 1886.

[Ann. Jard. Bot. Buitenzorg, 1me Série VII (1888), p. 213 seq].

Treub, curious to know whether a new vegetation was developing on Krakatao gathered information thereabout from sailors. He got different answers: according to some the thick layer of pumice was everywhere bare and sterile, but others affirmed having seen plants of some height here and there with the aid of telescopes. In order to get certainty on this point Treub availed himself of the opportunity offered by the renowned Krakatao-specialist Verbeek of visiting the island in the company of that savant and 8 other Europeans, none of them being a botanist. This first botanical exploration took place in June or July [1]) 1886, almost 3 years after the eruption.

Already when approaching Krakatao Treub observed numerous plants on the island, not only on and near the shore but also in the higher parts and even near the top. Unfortunately this observation did not induce him to extend his researches to those higher parts, where – as I have set forth above (p. 26) – part of the old vegetation might have survived and which therefore certainly ought to have been investigated. Consequently we do not know, and never shall know, whether in 1883 the vegetation was indeed entirely destroyed.

I mentioned already that Treub, feeling quite convinced of the total destruction of the vegetation, restricted his researches to a comparatively small part of the north-western side of the island. This locality consists of two rather different regions. The first is a level strip of beach, lying on the northern side of the rupture. Bounded on the north-western side by the basaltic rocks of Zwarte Hoek it extends in a south-eastern direction over a length of about 750 m. Towards the north-east it gradually slopes down to the sea; on the opposite side it is secluded from the interior by a steep wall of rock and pumice; on the southeast it terminates in a large talus of ashes, wich neither can be crossed nor ascended because

---

[1]) Treub himself gives *June*, Verbeek (by letters of April 25st 1922 and May 7th 1923) *July* as the month of the excursion. A photo of the party, taken at Zwarte Hoek, was published by Wing Easton in a paper dedicated to the memory of Dr. R. D. M. Verbeek, († April 9th 1926), Bijlage van het Jaarboek van het Geologisch Mijnbouwkundig Genootschap voor 1926. The legend to this photo wrongly gives the year 1883 for Treub's visit.

[2]) The voyage was made on the steam-vessel Soerabaja, commanded by skipper Mahne.

of the mobility of the loose materials and which bore as late as 1908 only a very scanty vegetation. The second region consists of the almost bare, easily accessible rocks of Zwarte Hoek and the steep slopes above it, the latter being ploughed by deep ravines and partly covered with ashes and pumice. Treub himself did not describe the locality investigated but Penzig 11 years afterwards gave the description quoted on pp. 29, 30. The rocky unweathered parts of the locality must have been (Cf. p. 30) very sterile; plants could only live in crevices and hollows and this was in the east-monsoon only possible because on Krakatao this monsoon, as a rule, is not strong, so that sheltered localities do not dry up much (Cf. p. 26, footnote 1).

From his finds Treub drew a number of conclusions which in consequence of the very defective knowledge extant in his time of the condition of life and the area of Indian plants, appeared quite correct and were universally accepted. But now they can be no more admitted to their full extent. Therefore I shall first discuss the plants found bij Treub, secondly the conclusions drawn by him, and in the third place I shall consider these conclusions in the light of our present knowledge.

Of the plants found by Treub unfortunately only the ferns, determined by Burck, are preserved in the Buitenzorg Herbarium. As to the higher plants it seems that only their names were noted down by Treub but that no specimens were collected or at least preserved, so that, for these, it is impossible to control the determination, a precaution which cannot be considered unnecessary for lists of plants dating from that time, as will appear from my discussion of the ferns. The fern-material has been preserved in a good condition but in consequence of the quite inadequate labelling it is of little or no value for oecological purposes: the labels only bear the note Krakatao or Zwarte Hoek; data about habitat, altitude above sea-level and frequency are entirely lacking. Neither in Treub's publication information on these points is found. Only by examining the conditions of life and the area of distribution in the Dutch Indies of the species found by him, one can get some rough idea of the composition of the fern-vegetation in the locality investigated in 1886; a detailed account is impossible. As to the higher plants of which the identity is partly not established the conditions are yet worse. Barring the general communication that they were scattered and few in number we know about those higher plants next to nothing except that for part of them the shore, for the rest the interior of the island is given as the habitat and even these statements are sometimes liable to suspicion. Treub whose doubtless extraordinary merits lie neither in the domain of oecology nor in that of phytogeography has worked here in a field where he was no expert. Therefore he did not conduct the investigations regarding the new flora in a scientific way so as to lay a base for all subsequent researches. This however has largely to be ascribed to the very unfavourable circumstances under which he worked.

On the beach Treub found a number of fruits and seeds which were not collected or at least have not been preserved. According to his paper they belonged to the following species:

1. **Heritiera littoralis** Dryand. (1 specimen).

A tree which is rather common in Java especially in the older, less saline parts of tidal forests on clayey soils [1]), and the localities behind, but which sometimes is found growing immediately behind the flood-mark. The reproduction takes place exclusively by fruits [2]) which are rather often washed ashore. But only where the locality is humid, young plants develop; elsewhere either the seeds do not germinate or the seedlings soon perish [3]). Plants of *Heritiera* have, to my knowledge, not yet been found on Krakatao.

1) One should well distinguish tidal forests on clayey soil from those on coral-reefs. The latter are, on the whole, poorer in species but possess some plants (*Pemphis acidula, Rhizophora stylosa*) which do not occur on clay, whilst in tidal forests on coral-reefs *Sonneratia acida*, — *Rhizophora mucronata*, — *Finlaysonia maritima* and *Acrostichum aureum* are always lacking, — *Rhizophora conjugata* is found as well on a clayey soil as on coral-reefs [Cf. also Schimper, Indo-Malayische Strandflora (1891), p. 185 seq]. On humid sandy beaches rich in shells (as near Maroonda, East of Tandjong Priok) mangrove-plants may occur which, for the rest, are only found on coral-reefs.—It is perhaps not superfluous to observe here that in the latest 25 years in several localities in Java the composition of mangrove-forests has been greatly modified by the reckless cutting out by the natives of species of economic value (piles, timber, firewood, tanning bark). Almost everywhere in the neighbourhood of large settlements where the Forestry Service has not been able to intervene, only the species of little or no economic value are left [4]). And not only in the tidal forests but almost everywhere in the lower and middle regions of Java intact primeval forests have become exceedingly rare. What by the non-expert is taken for primeval forest is, as a rule, old secondary forest.

2) Described by Boerlage in Ann. Jard. Bot. Buitenzorg Suppl. II (1898), p. 137; Schimper, Indo-Malayische Strandflora (1891), pp. 164 en 184, tab. VII fig. 2 and Massart, Un Botaniste en Malaisie (1895), p. 298 (150).

3) Cf. also Schimper, Indo-malayische Strandflora (1891), p. 162: „Keimlinge von Heritiera findet man" (on the sandy beach of Tjilatjap) „im Verhältniss zu der enormen Menge der angeschwemmten Früchte wenig".

4) Laudable exceptions to this rule may be met with near fisher-villages, where species of *Rhizophora* are often planted by the natives who use the bark for tanning their nets, the wood for fuel and other purposes. Cf. Heyne, De nuttige planten van Nederlandsch-Indië (Useful plants of the Dutch Indies), 2' druk, II (1927), pp. 1162 seq.

2. Terminalia catappa L. (2 specimens).

A tree which in Java is common on sandy beaches and also occurs along the lower course of some rivers. The reproduction takes exclusively place by fruits [1]) which are very often washed ashore [2]). The low beaches of Krakatao are very favorable to this tree, which was found there on many excursions.

3. Cocos nucifera L. [3]) (1 specimen).

This palm is extensively cultivated in the Dutch East-Indies, especially in the coastal regions and the lower plains. On the beach one often finds its fruits, which, as a rule, are either unripe or so much damaged as to be unfit to germinate [4]). It seems that Treub did not examine whether the coco-nut found by him had preserved its germinative power. Eleven years afterwards Penzig [5]) found no coco-palms in the same locality. In 1906 Ernst and his companions found near Zwarte Hoek some young specimens. In 1908 these specimens had disappeared but I found a few very young trees rather far behind the flood-mark. If these specimens had remained in life they would in 1919 have been rather large and have fructified. But Docters van Leeuwen mentions [6]) to have found there only a few young specimens, and these stood more to the south-east, so that they can not have been the same as those of 1908. It is generally believed but was never proven that fruits of this species are carried over long distances by ocean-currents *without losing their germinative power* so that by this means the plant can establish itself in the locality reached. But it is certain that man takes a very active part in its distribution. *The numerous coco-palms on the coral islands in the Java-sea and along the coast of Java* [7]) *are all planted specimens.*

1) Described by Schimper Indo-Malayische Strandflora (1891), p. 170, tab. VII, fig. 11.

2) Guppy found them washed ashore on the Keeling-islands.

3) In Treub's paper by misprint **suceifera.**

4) Cf. Schimper, Indo-Malayische Strandflora (1891). pp. 161, 162.- Ernst. [Neue Flora Vulkaninsel Krakatau (1907), p. 28] also declares that the coco-nuts found in 1906 on the beach of Krakatao were „von Tieren siebachtig durchbohrt oder völlig ausgehöhlt".

5) Ernst, Neue Flora Vulkaninsel Krakatau (1907) 16.

6) Ann. Jard. Botanique Buitenzorg XXXI (1919). 111.

7) Koorders (Exkursionsflora I, 246) mentions that the found undoubtedly wild-growing coco-palms in East-Java but he does not give any other reason for his belief than that the fruits of the trees in question were small and thick-walled. This form of coco-nut, he says, is never cultivated. But it is already mentioned in Miquel, Flora Indiae Batavae III (1855), p. 71 under the cultivated forms of Cocos *nucifera* as var. *cistiformis* and recently it has also been found cultivated in Sumatra. The shell of this variety is often used for native carvingwork. Cf. Heyne, Nuttige Planten Ned. Indië, 2e druk, I (1927) p. 411 sub var. *cystiformis.*

Though, generally speaking, it is difficult or impossible to make out whether a coco-nut, carried by the sea to the shore of an island, originates from this same island or from another one, I admit the possibility that on Krakatao, from time to time, living coco-nuts coming from other islands may be washed ashore. But, by this, the species has by no means established itself in the locality reached. If circumstances are most favourable the fruit germinates immediately behind the floodmark where the seedling is exposed to numerous dangers, except on rapidly increasing beaches. Here and there on Krakatao and the neighbouring islands many specimens, even groves, of this palm have been found but it is sure that many, if not all, of these were planted by man. It is a well-known fact, also memorated by Docters van Leeuwen [1]), that when fishermen for some purpose go ashore on the uninhabited islands of the Java-sea and find a young seedling of a coco-palm on the beach, they remove it from its unsafe habitat and plant it in the interior. They do so, of course, with a view to the future fruits, which may supply them with drink or food or form an article of commerce. For this same purpose germinating coco-nuts are sometimes introduced from elsewhere. That coco-palms had been planted on an uninhabited island I have observed myself in 1921 on the coral-island Klein Kombuis (afterwards declared a Nature monument) where, in the midst of the dense forest covering the entire island, plots had been cleared for this purpose. Something like this has also happened on Krakatao: not only the grouping of some of the coco-palms there points to human influence, *but it has appeared from an investigation instituted some years ago by the Resident of the Lampong Districts* (to which Krakatao belongs) *that after the eruption one Hadji Djoemali* (of Kalianda) *has planted coco-palms on the island* [2]). Therefore I think it quite possible that all coco-palms occurring on Krakatao are either planted specimens [3]) or form the offspring of these, growing quite near their parent.

In the Dutch Indies the fact that fruits of the coco-palm reach, without human assistance, a safe locality to germinate in is of so extremely rare occurrence, that spontaneous migration from one island to another, or even from one point of the shore to another

---

1) Ann. Jard. Bot. Buitenzorg XXXII (1923), 142.

2) Tectona XVIII (1925), p. 262.

3) In 1906 Ernst and his companions found a group of coco-palms more than 50 m. from the shore and *above flood-mark*.– Docters van Leeuwen [Ann. Jard. Bot. Buitenzorg XXXIII (1923) 142] found on Verlaten Eiland several coco-palms far from the sea, without „the possibility of explaining this by the supposition that the coast-line had shifted".

somewhat remote one, may be considered as practically not happening at all. During the latest 20 years I have travelled on foot over a distance of several hundreds of kilometers along the shores of Java and the smaller islands in the vicinity and never found there a single coco-nut which after having been washed ashore, had germinated and rooted. Schimper when investigating the beach near Tjilatjap (on the southern coast of Java) made the same experience as myself; he expressly states [1]): „Junge Cocospalmen findet man gar nicht" and I mentioned already that the coco-nuts, which Ernst [2]) and his companions found washed ashore on Krakatao, were pierced like a sieve and quite empty. Ernst has figured [3]) a rather large coco-palm growing on the south-eastern side of Krakatao in a spot where the beach was decreasing. According to the legend this tree was growing on the beach behind the high-tide mark but the figure itself shows that it was already attacked by the sea and therefore every moment liable to be destroyed. This tree might quite well have been planted at some distance from the sea which may have reached it by washing away the beach: a shore so much inclined as the one figured in the left half of the plate is only found in localities where the beach decreases.

As a proof that Cocos *nucifera* can be spread by sea-currents Docters van Leeuwen mentions [4]) the fact, that he found on the (inhabited) island Sebesi (north of Krakatao) „some big „coco-nut plants growing on the mostly inaccessible cliff-walls a few „yards above the sea, probably cast up there by the high waves. Close „to the water and higher up not a single coco-palm was to be seen". But cliff-walls inaccessible to Europeans navigating on the sea below need not at all be so to nimble-footed natives coming from the interior, and therefore the argument advanced by Docters van Leeuwen is not conclusive. His assertion that the coast is (and was?) *never* visited, should he confirmed, if possible, by a minute investigation in loco before it may be taken for correct. I shall revert to his paper on Sebesi in Chapter XII.

How dangerous it is to eliminate man when accounting for the presence of coco-palms in uninhabited islands appears from the following communication of W. Botting Hemsley in Nature, April 10 1890, p. 537: „The question whether the coco-nut palm is

1) Indo-Malayische Strandflora (1891), pp. 161, 162.
2) Neue Flora Vulkaninsel Krakatau (1907), p. 28.
3) Neue Flora Vulkaninsel Krakatau (1907), tab. V, fig. 8.
4) Ann. Jard. Bot. Buitenzorg XXXII (1923), p. 142.

„capable of establishing itself on oceanic islands, or other shores for „the matter of that, from seeds cast ashore, was long doubted ..... „There is now *absolutely incontroversible* [1]) evidence, that it is „capable of doing so even under apparently very unfavourable cir- „cumstances ..... In the last part of the Proceedings of the Royal „Geographical Society" (March 1890, p. 159) „Mr. J. J. Links gives „an account of the natural history of the (newly raised) Falcon „Island. From this interesting contribution to the source of insular „floras we learn that he found two young coco-nut palms, not in a „very flourishing condition, it is true, but they were there and *had „evidently obtained a footing unaided by man* [1])".

Alas. A fortnight afterwards in the same periodical a note, inserted by W. J. L. Wharton, appeared saying: „With reference to „Mr. Hemsley's note .... I regret to inform him that the two young „palms found on Falcon Island were placed there by a Tonga chief „of Namuka who in 1887 had the curiosity to visit the new-born „island and took some coco-nuts with him. This information I received „from Commander Oldham, who had been much interested at finding „these sprouting nuts at some 12 feet above sea-level and well in „from the shore of the island, but who found out the unexpected „facts in time to save me of making a speculation somewhat similar „to Mr. Hemsley's.

Prain [2]) also, when describing the vegetation of the islands of the coco-group (north of the Andamans) does not suppose that the coco-palm, which is frequent there, has established itself without the intervention of man. He says (l.c. p. 332) „*Cocos nucifera is* extremely „abundant in all the islands. Probably it is not truly indigenous „though perhaps not intentionally introduced. Once established the „species spreads with great rapidity. It should not be forgotten that „at some remote period a colony may have been started in the „Coco-group and then abandoned. It is known that in recent times „two such attempts have been made". And further (l.c. p. 375): „It „seems that for this particular group of islands, although the spread „of coco-nuts within the group is undoubtedly due to the agency „of the sea, the ocean current theory does not explain the presence „of the species and that the original introduction has more probably „been due to human inference".

[1]) Italics by me. The two communications cited here were already referred to by Treub in Teysmannia I (1890), p. 239.

[2]) Vegetation of the Coco-group in Journal As. Society Bengal, LX, part 2, pp. 283 seq.

I am very sceptical as to the possibility of the coco-palm being spread by the sea over large distances, which has not been proven by a single reliable observation. It is impossible to make out, without minute investigations, whether a tree, apparently growing spontaneously, was not planted originally near a settlement that was afterwards deserted by man and swallowed up by jungle. In the tropics such often happens in thinly populated regions.

4. **Pandanus** spec.

On and near the Dutch East-Indian beaches occur, as far as is known at present, 7 species of Pandanus, most commonly the very polymorphous *P. tectorius Sol.*, further *P. andamanensium Kurz,* – *P. bidur Jungh.,* – *P. dubius Spreng,* – *P. labyrinthicus Kurz,* – *P. polycephalus Lamk.* and *P. nitidus Kurz,* the last one but rarely. The dispersal of these species is brought about by the fruits, which may be carried along either by water (sea-currents, tidal streams) or by animals. It has not been proven that animals carry the fruits from one island to another. It seems that Treub, who mentioned this genus twice in the record of the fruits found by him, has found the fruits of 2 species of Pandanus (or fruits which he meant to belong to 2 species), both in a single specimen. All plants of Pandanus found until now on Krakatao belonged to *Pandanus tectorius Sol.* (P. littoralis Jungh.). In the western part of the East Indian Archipelago this is by far the most common species of the genus on sandy beaches; its fruits [1]) are very often found washed ashore and seedlings also are quite common.

5. **Barringtonia speciosa** Forst. (5 specimens).

This species, at the present time named *B. asiatica Kurz,* is a tree of sandy, especially coralligenous or calcareous beaches, locally often occurring in great numbers but which may also be entirely lacking, even in the formation named by Schimper after it. As a rule it is in Java much less common than *Terminalia catappa L.* The reproduction is effected exclusively by fruits [2]) which possess floating power and are frequently found washed ashore. The beaches of Krakatao form a very fit habitat for this species which was found there at several excursions.

---

[1]) Described by Schimper, Indo-Malayische Strandflora (1891), p. 177, tab. VII, fig. 12a, 12b.

[2]) Described by Schimper. Indo-Malayische Strandflora (1891), pp. 172 and 184, tab. VII, fig. 3 under the name of *B. speciosa.*

6. **Calophyllum inophyllum** L. (3 specimens).

A very common tree of sandy beaches. It reproduces itself only by fruits [1]) which possess floating power and are very often found washed ashore. The beaches of Krakatao are very favourable to this tree, which was found there at several excursions.

As far as the identity of the fruits and seeds, found by Treub on the shore of Krakatao, is established they belong to such trees as are frequent on the beaches of the Dutch East Indies. Though, as a rule, seeds found washed ashore on an island may quite as well originate from that island itself as from any other one, we have to admit in the present case introduction from elsewhere, as it was impossible that within 3 years after the eruption fruiting specimens of the trees in question could have grown on Krakatao.

But from the fact that these fruits and seeds had reached the shore, even when assuming that they had retained their germinative power, may by no means be inferred, that the species they belonged to had succeeded in establishing themselves on Krakatao. In order to grow up to adult plants the seeds must come to a rest and therefore reach the locality *behind the flood-mark*, which, circumstances being favourable, in the case of small seeds may be effected by the waves and by the wind but in the case of big ones only by a very high flood or a high wave and occasionally perhaps by animals. When the tide runs low, seeds are often left lying in and before the flood-mark, but when shortly afterwards the tide is high, these seeds for a great part may become floating again. Where the beach rapidly decreases and therefore forms a steep slope towards the sea, by far the greater part of the seeds do not reach a safe locality behind the flood-mark; they remain at the foot of the steep declivity and, when the tide is high, are in continual movement so that they cannot germinate or, at least, not root and grow [2]); all these seeds perish. The few seeds carried on such beaches behind the floodmark by the aid of a high wave, may germinate, but the young plants are far from safe as not only every next high wave may damage or kill them but also the coastal line is shifting backwards and the habitat, in course of time, is destroyed.

The best chance to live have fast growing plants possessing long creeping stems, rhizomes or stolons by which they can soon get away from their unsafe first habitat. Sometimes a beach increases

[1]) Described by Schimper, Indo-Malayische Strandflora (1891), p. 175 and 184, tab. VII, fig. 9. Guppy found them in the drift-zone on the Keeling Islands.

[2]) Cf. also Kamerling in Natuurk. Tijdschr. Ned. Indië LXXXI (1921), p. 22 seq.

during a season and decreases during the next so that plants that have lived some time under favourable conditions afterwards perish, their habitat being destroyed. Only on increasing or stable beaches circumstances continuously favour germination, at least in the west-monsoon, but even there many seedlings perish, the substratum being often unfit for their further growth [1]). Along the sea-side of the Barringtonia-formation one regularly finds seedlings of *Nipa fruticans Wurmb,* and *Avicennia marina Vierh.* but these, without exception, die early as they can only live in a constantly marshy or at least very humid soil [2]). The fruits of *Gluta renghas L.*, a tree chiefly occurring on riversides and in bog-forests in the plains, are frequently washed ashore but young plants of some size are on the beach found only in such localities where the ground-water is sweet and the water-table high. Of *Pangium edule Reinw.* one finds on the beach often seeds but never seedlings [3]) and the same is the case with sundry species of *Quercus.* Schimper [4]) mentions the same fact for the seeds of *Chydenanthus excelsa Miers.* The fruits of *Cerbera manghas* are very often found lying before or even on the beach-wall; they may germinate there but soon afterwards perish. In 1921 Kamerling [5]) quite rightly observed that on a beach where many seeds or fruits of a given species are found, very often not a single well-developed seedling grows except quite near an adult specimen of the same species, the fruits or seeds of which therefore may have been transferred over some slight distance only, but need not at all have come over from far away. It is beyond doubt that, except in well-sheltered localities as f.i. the increasing lee-side of coral-islands in the Java-sea [6]), only an exceedingly small part of

1) Cf. also Schimper, Indo-Malayische Strandflora (1891) p. 186 seq. The coral-island Pulu Lang in the Java-sea mentioned by him should not be confounded with the volcanic island of the same name [7]) in the Sunda-straits. Schimper (l.c. p. 183) draws also attention to *Lodoicea Seychellarum* „deren Früchte bekanntlich ganz regelmässig an der Küste der Lakediven und sogar von Sumatra angeschwemmt werden und dennoch die Palme nicht über die Seychellen hinaus verbreitet haben".

2) On the sand- and pumice-beaches of Krakatao and Verlaten Eiland Ernst and his companions found in 1906 many young seedlings of *Nipa fruticans Wurmb,* partly still living, partly already dead, but not a single older specimen (Cf. Ernst, Neue Flora Vulkaninsel Krakatau (1907), p. 38).

3) This was already mentioned by J. Massart [Un botaniste en Malaisie p. 299 (151)]: „les rivières les ont entraînées vers la mer, les flots les ont rejetées à la côte, mais sans aucun profit pour l'espèce car celles ont été tuées par l'eau salée".

4) Schimper, Indo-Malayische Strandflora (1891), p. 173.

5) Kamerling in Natuurk. Tijdschr. Ned.-Indië LXXXI (1921), p. 22.

6) The vegetation of such parts of the beach of coral-islands as are exposed to the surf often differs much from that of the sheltered parts. In the western half of the Java-sea the south and sout-eastern sides of the coral islands bear as a rule, a *Pes-caprae* or a *Barringtonia* formation, the north- and north-western sides a coral-reef mangrove (Cf. pag. 38, footnote 1).

7) *Pulu* is the malayan word for *island,* Dutch: *eiland.*

the seeds and fruits washed ashore contribute to lastingly enrich the flora.

Everyone who is, from personal view, acquainted with the drift-zone will admit that on the beach examined by Treub a great number of small seeds and fruits may have been present that remained unobserved [1]). For persons who never saw tropical drift-zones a look at the fine picture given by Ernst [2]) will suffice.

On the shore of Krakatao Treub found young specimens of:

1. **Erythrina** spec.

Until now 2 species of *Erythrina* have been found on the beaches of Krakatao. The first of these, *E. variegata L.* var. *orientalis Merr.* (E. indica Lamk.) is in Java a rather common tree of sandy beaches; in the interior it is often cultivated. Its indehiscent pods are often found washed ashore [3]); like the seeds they possess floating power. The second species, *E. fusca Lour.* (E. ovalifolia Roxb.) is an inhabitant of riversides in the lowland plains. Its pods burst when ripe; the seeds can float and are often carried by the rivers to the sea. They frequently reach the shore but develop into trees only where the sweet ground-water rises near the surface and consequently may be easily reached by the roots of the seedlings. It is impossible to decide which of these two species was found by Treub; neither of them was eleven years afterwards found back by Penzig in the same locality. But in 1906 and 1908 only the first was seen at Zwarte Hoek, so that it may have been this species that Treub collected.

2. **Calophyllum inophyllum** L. See p. 44, No. 6.

This species was not found back by Penzig in 1897 in the same locality. Like the former it illustrates what I have said (p. 44) about the dangers to which young seedlings on the beach are exposed.

3. **Cerbera odollam** Gaertn.

This species, at the present time named C. *manghas L.* is a small tree locally frequent on clayey soils along tidal creeks and

---

1) Schimper [Indo-Malayische Strandflora (1891), p. 172] already pointed out that in the drift-zone small fruits may easily escape the eye. The record (l.c. 160) given by him of seeds and fruits washed ashore contains the names of many plants not mentioned by Treub, but the seeds or fruits of which most probably land from time to time on Krakatao.

2) Neue Flora Vulkaninsel Krakatau (1907), plate IV, fig. 6. This increasing beach should be compared with the decreasing one pictured on plate V, fig. 8, left side,

3) Guppy found them on the Keeling-islands in the drift-zone.

mouths of rivers. On the beach this tree grows only where sweet or but slightly brackish ground-water rises near the surface [1]). On Krakatao it has repeatedly been found, i.a. in 1897 by Penzig, in 1906 by Ernst, in 1908 by myself, in 1919 by members of the Congress of Natural Science. The specimens found in 1908 were all very young; this proves once more that young seedlings on the beach often do not grow up into trees. The reproduction is effected exclusively by fruits [2]) which are exceedingly fit to be carried by water and are very frequently found washed ashore. The specimens of this tree which are found in the interior of Java are without exception either cultivated or the offspring of the latter, growing quite near.

4. Hernandia sonora L.

This name has been given to different species. The one found by Treub was probably *Hernandia peltata Meissn.*, a tree which in the western half of the Dutch Indian Archipelago is rather common on sandy beaches [3]). The beaches of Krakatao are a fit habitat for this species which was found there at several excursions. The reproduction is effected exclusively by the fruits which possess floating power [4]) and are often found washed ashore [5]).

5 and 6. Two **Cyperaceae.**

This vague indication is of no value for the study of the renewal of the flora. The Cyperaceae which until now were found on the north-western side of Krakatao on or near the beach were, as far as I know, *Cyperus pennatus Lmk.*, *Fimbristylis spathacea Roth* and *Remirea maritima Aubl.* This last species, occurs only on young beaches consisting of very loose sands and possesses a long creeping rhizome which, at short distances, emits young individuals, whilst the other two species grow in tussocks and occur not only on the beach but also rather far behind it; they prefer a less permeable substratum. The fruits of these 3 species can be spread by water. It is needless to discuss the Cyperaceae found by

1) Where this is not the case (as on the coral-island Pulu Lang in the Java-sea, investigated by Schimper) either the seeds do not germinate or the seedlings die early [Schimper. Indo-Malayische Strandflora (1891), p. 190].

2) Described by Schimper, Indo-Malayische Strandflora (1891), p. 168 tab. VII fig. 22 under the name of *Cerbera odollam Gaertn.*

3) The only other species of *Hernandia* occurring in the East Indian Archipelago, *H. ovigera* L., is no littoral tree [Cf. Koorders en Valeton, Bijdrage VII, p. 110 and Rumphius, Herb. Amboinense II, 257. tab. 85 (Arbor regis).

4) Described by Schimper, Indo-Malayische Strandflora (1891), pp. 176. 177. tab. VII.

5) Guppy found them in the drift-zone on the Keeling-islands.

T r e u b, because their identity is not established and other species of the same family (*Kyllinga monocephala Rottb.,* – *Cyperus kyllingoides Vahl,* – *C. stoloniferus Retz*) may have occurred on the beach. *Cyperus pennatus* and *Fimbristylis spathacca* were in 1897 and in 1906 found on the beach near Zwarte Hoek.

7. **Ipomoea pes caprae** L.

A plant with very long (up to 35 m.) creeping stems, in the Dutch Indies of common occurrence on sandy beaches, *but which is also found in many localities in the interior, far remote from the sea.* The reproduction is effected by ramification of the creeping stems and by seeds. The latter possess floating power, because the testa is not entirely filled up by the germ [1]); they are also spread by rain-water flowing rapidly down. The beaches of Krakatao offer a very favourable habitat to this species, which was found there at every excursion.

8. **Gymnothrix elegans** Büse.

At the present time named *Pennisetum macrostachyum Brongn,* a robust, tillering grass, without stolons. The reproduction is effected by fruits which, when ripe, remain enclosed in the spiculae. The latter fall off as a whole and bear at their bases an involucre of very long bristles by the aid of which they may be easily spread over some distance by the wind. The statement of T r e u b that this grass is common in Java should be understood in this way that the plant is often cultivated there for ornamental purposes; spontaneously growing it has been found in the island only a few times. It is restricted to sunny, rather sterile localities; hence it found in 1886 and yet much later, on Krakatao many localities fit to grow in. In 1906 it occurred in some specimens in crevices of the rocks at Zwarte Hoek, on the same side of Krakatao where T r e u b had discovered it 23 years before. It is no littoral plant and, if T r e u b did really find it on the shore *in the oecological sense of the word* it must have been in a single specimen gone astray from the interior where T r e u b likewise found the plant. But probably he comprised under „shore" also the quite leached-out older parts of the increasing sandy beach at the base of the steep wall (Cf. p. 36) by which it is secluded from the interior. There or on the wall itself *Pennisetum* may have grown then as it did in later years.

---

1) Cf. S c h i m p e r, Indo-Malayische Strandflora (1891), p. 163.

9. Scaevola Koenigii Vahl.

At present named *Sc. frutescens Krause*. This is a shrub which in the Dutch East Indies is very common on sandy beaches lying either on coral or on another subsoil. The fruits [1]) possess a fleshy pericarp that soon becomes dry, and a ribbed endocarp developed into a floating tissue; they are frequently found washed ashore [2]). The beaches of Krakatao are very fit for this species, which was found there at every excursion.

As far as the identity of the plants found by Treub on the shore is well-established, they all belong to species which are frequent on the beaches of the Dutch East Indies, with the sole exception of *Pennisetum macrostachyum*, an inhabitant of sunny sterile localities, such as the rocky slopes, bare or covered with ashes or pumice, which lie near the part of the beach investigated by Treub. This beach did not exist before the eruption. As we may certainly assume that by the eruption of 1883 the original littoral vegetation of Krakatao was entirely destroyed, it is very probable that all the true littoral plants found by Treub were introduced from elsewhere.

For the interior, „la montagne proprement dite", Treub mentions the following species:

PHANEROGAMS.

1. Wollastonia spec.

Wollastonia, at present reduced as a section to the genus *Wedelia*, is represented in the Dutch East Indies by a small number of species of which only one, *Wedelia biflora D. C.* (Wollastonia glabrata D. C.), has been found until now on Krakatao. To this species belong also, as appears from materials preserved in the Buitenzorg Herbarium, the plants mentioned for Krakatao in the list of Penzig [3]) under the names *Wedelia asperrima Boerl.* and *Wedelia scabriuscula Boerl.* Hence it is probable that the plant recorded by Treub as *Wollastonia*, was also *Wedelia biflora*.

Wedelia biflora is a very common halophyte on the Indian beaches and the level or almost level localities [4]) immediately or at some small distance behind these. Often also it is found along the mouths

[1]) Described by Schimper, Indo-Malayische Strandflora (1891), p. 172, tab VII, fig. 25, under the name of *Sc. Koenigii*.

[2]) Guppy found them in the drift-zone on the Keeling Islands.

[3]) Ann. Jard. Bot. Buitenzorg XVIII (1902) 105.

[4]) Cf. J. Massart, Un Botaniste en Malaisie (1895), p. 290 (142) (under the name of *Wollastonia glabrata D.C.*); p. 294 (146) (under the name of *Wollastonia*).

of rivers up to the point where the influence of the tide becomes perceptible. Its fruits lack a pappus; the pericarp is developed into a floating tissue [1]); in a 3 % solution of NaCl they remained floating for more than 20 weeks. They are chiefly dispersed by water; by wind they might in bare localities be driven forth a few meters but on account of their weight they are unfit for a somewhat prolonged flight through the air. It is quite impossible that they have crossed over to Krakatao from some of the neigbouring islands „auf den Flügeln des Windes" as Penzig [2]) has it.

If Treub has really found this plant „sur la montagne proprement dite" and not at a short distance from the beach in a low-lying ravine opening on the coast, with perhaps a saline soil [3]), it can have occurred only in a few specimens gone astray from the shore. This species cannot possibly have taken an important part in the vegetation of the higher interior parts.

2 and **3.** Two species of **Conyza.**

The genus *Conyza,* as delimited by Hoffmann in Engler und Prantl (Natürl. Pflanzenfamilien IV, pars 5, 169) is in the Dutch East Indies only represented by a few species in the mountainous districts above 1000 m. Treub cannot have found these on the lower parts of Krakatao. But in the only flora of the Dutch Indies Treub had at his disposal (Miquel, Flora Ind. Bat. II, 40) *Conyza* includes also the present genera *Pluchea* and *Blumea,* whilst a species of *Erigeron* so closely resembles *Conyza* that it is often wrongly determined as such. In these 3 genera we have to look for Treub's species of *Conyza.*

Pluchea is in the Dutch Indies represented only by *Pl. indica Less.* (= Conyza indica Miq.). This is a halo-xerophilous shrub of sunny or slightly shadowed, periodically very dry and frequently saline localities, chiefly occurring on the heavier soils and in stony or calcareous localities either near the sea or rather far from it; it is also found around salt-water springs. It is not at all a beach plant, though sometimes recorded as such. In the interior it is often cultivated in

1) Cf. Schimper, Indo-Malayische Strandflora (1891), p. 172, under the name of *Wollastonia glabra;* p. 157, under the name of *Wedelia biflora.*

2) Cf. Penzig in Ann. Jard. Bot. Buitenzorg XVIII (1902) p. 109.

3) Verbeek, Krakatau, Dutch Edition (1888), p. 302 (translated): „The eruptive „products impregnated with sea-water were (in Oct. 1883) in the lower layers still very „hot; the sea-water and especially the rain-water oozed through the porous pumice-sand, „carried the dissolved salts of the sea-water downwards and deposited them as a white „or gray salt-crust in the localities were the hot water could escape". The eruptive „products contained according to Verbeek [Krakatau, Dutch Edition (1888), p. 292 (analyses 7, 8, 9)] 0,65 – 1,09 % salt.

living fences or for culinar or medicinal purposes. It does not blossom before it is somewhat advanced in age; the fruits which are plentifully produced possess a pappus. In Java the plant is very common and blossoms and fruits throughout the year. In 1897 it was found on the north-western side of Krakatao where some localities are not unfit for it; P e n z i g however does not mention it under the common species. I do not think it probable that T r e u b has found this plant, so easily recognizable by its leaves and its peculiar smell, without being able to find out its name. Every native accompanying him [1]) must have known it.

Of the genus *Blumea*, all species of which possess a pappus, until now the following species have been found on Krakatao:

Blumea lacera D. C. (Conyza lacera Burm.), an annual herb of sunny roadsides, alangfields and other dry grassy localities and neglected fields, occurs throughout Java from sea-level up to ± 1000 m. In the surroundings of Batavia and Tandjong Priok this species is very common; it was found with certainty on Krakatao for the first time in 1908. I do not think it very probable that T r e u b has found this easily recognizable plant which blossoms already when young, without being able to find out its name.

Blumea balsamifera D. C. (Conyza balsamifera L.) a robust, often shrubby plant of sunny localities, alangfields, grass-jungles and light woods, occurs throughout Java from the plains up to ± 2000 m. In the surroundings of Batavia it is not rarely found spontaneously growing but more often yet it is cultivated by the natives for its medicinal properties. In 1897 it was found by P e n z i g on the north-western side of Krakatao. The plant does not require a very fertile soil; it blossoms only when somewhat advanced in age. I think it quite possible that one of T r e u b's species of *Conyza* was this *Blumea* occurring in specimens too young for recognition.

Erigeron linifolius Willd, wrongly named by B o e r l a g e *Blumea hieracifolia D. C.* and mentioned under this name in the list of P e n z i g, was found in 1897 on Krakatao. This species, which like all the rest of the genus possesses a pappus on the fruits, deceivingly resembles a *Conyza* and has been described as a species of that genus (C. angustifolia Thwaites). As late as 1905 it was lying under

[1]) As appears from the photo mentioned on p. 36, footnote 1, that party was accompanied by at least 5 natives.

this name in the Buitenzorg Herbarium. This herb is not yet mentioned in the Flora of Miquel and could therefore not be determined by Treub or Burck. Probably it was introduced into Java about the middle of the 19th century [1]), possibly from tropical America; at present it is spread throughout the island from the plains up to an altitude of ± 2650 m. It chiefly occurs in sunny or slightly shadowed fields and does not require a fertile soil; it often grows gregariously on exhausted or washed-off soils or on vertical banks clothed with kremnophytes. In the surroundings of Batavia it is rather common in several localities. I think it quite possible that the second *Conyza* of Treub was identical with this plant.

4. Senecio spec.

The only species of *Senecio* [2]) occurring in the lower regions of the Dutch East Indies is the violet-flowered *Senecio sonchifolius Moench.* This is in Java from the plains up to an altitude of ± 1750 m. in many localities a rather common weed of fields and road-sides; in 1897 it was collected on the north-western side of Krakatao. This species, recorded by De Candolle (Prodromus VI, 302) and by Miquel (Flora Ind. Bat. II, 101) under the genus *Emilia,* may have been found by Treub in 1886 but I do not think so. *Senecio sonchifolius* so little resembles the European species of the genus, some of which were doubtless known to Treub, that I cannot believe that the latter, who was no systematist and had to rely on Miquel, reckoned it to *Senecio,* contrary to the opinion of De Candolle and Miquel.— I think it much more probable that he has found *Erechthites hieracifolia Rafin.,* an American weed of agriculture introduced already long ago [1]) and at present frequent in many not too dry localities of Java and Sumatra, from the plains up to an altitude of 2100 m. but not yet mentioned in Miquel's flora. It somewhat resembles *Senecio* and was collected in 1906 at Zwarte Hoek, the same side of the island as investigated by Treub.

Both *Senecio sonchifolius* and *Erechthites hieracifolia* occur in the surroundings of Batavia and Tandjong Priok. The fruits of both possess

[1]) This now very common plant is not mentioned by Junghuhn and seems not to have been found by any of the older collectors. As far as I know Otto Kuntze was the first who found it in Java (1875).

[2]) The plant mentioned by Koorders (Exkursionsflora III, 345) for the lower regions of Java under the name of *Senecio tenuifolius Burm.* does not belong to this genus but is a *Chrysanthellum.* The statement on the same page that *Senecio araneosus D. C.* occurs in the plains of Java is incorrect; this species is restricted to the mountainous zone above 1000 m. as is the introduced *Senecio vulgaris L.* Other species of the genus have not yet been found in Java.

a well-developed pappus; both may be found in sterile and in fertile localities, though they prefer — especially *Erechthites* — the latter.

5. **Scaevola Koenigii** Vahl (Sc. frutescens Krause). See p. 49, No. 9.

In Java this species is restricted to the beach where it is often found in great numbers. If Treub has really found it in the interior of Krakatao [1]), we have to apply also to this species what I have said on p. 50 about the occurrence of *Wedelia*.

6. **Gymnothrix elegans** Büse (Pennisetum macrostachyum Brongn.). See p. 48, No. 8.

7. **Phragmites Roxburghii** Steud.

At the present time named *Phragmites Karka Roxb.* This is a meso-hygrophyte, which has not yet been found with certainty on Krakatao. The plant found by Treub was probably *Neyraudia madagascariensis Hook. fil.* var. *Zollingeri Hook. fil.* (Phragmites Zollingeri Steud.) which much resembles *Phragmites* but may easily be distinguished by its massive culms and villous glumes. As appears from materials preserved in the Buitenzorg Herbarium *Neyraudia* was found in September 1896 by Boerlage on one of the other islands of the Krakatao-group and named by him *Phragmites Roxburghii*. *Neyraudia* is a grass occurring in sunny and usually rather sterile localities, especially in stony places [2]). Consequently in 1886 and yet much later the plant may have found on Krakatao many habitats quite fit for it. Elsewhere on Krakatao this species seems not yet to have been observed. The statement of Ernst [3]) that in 1906 the vegetation on the eastern slopes of the island was partly composed of *Phragmites Roxburghii* must be due to some lapsus memoriae, perhaps brought about by his having read the paper of Penzig [4]);

[1]) Schimper [Indo-Malayische Strandflora (1891), p. 185] pointed already out how uncommon the habitat given by Treub was – Koorders [Teysmannia IX (1899), p 272] rightly mentions as a great exception that he found once *Scaevola frutescens* in a single specimen upwards of 30 m. above sealevel on the almost bare brick walls of a deserted fortress on the beach of Karang Bolong on Nusa Kambangan, which for the greater part was overgrown by jungle. The numberless other specimens of this species, observed by him in Java, Celebes and Sumatra, grew on the shore; I also found the plant only on or quite near the beach and there often in great numbers.

[2]) The plant was found i.a. by Mr. A. E. Kerkhoven here and there on the very stony lower slopes of Mount Guntur. This material which is preserved in the Buitenzorg Herbarium was determined by Boerlage, with some doubt, as *Phragmites* [Cf. Teysmannia VIII (1898), p. 506]

[3]) Neue Flora Vulkaninsel Krakatau (1907), p. 32.

[4]) Penzig in Ann. Jard. Bot. Buitenzorg XVIII (1902), p. 99.

in 1906 neither *Phragmites* nor *Neyraudia* were found on Krakatao and in the record of the plants found by him and his companions on the island Ernst [1]) did not mention these grasses for his trip. Nor were they, to my knowledge, found at any of the subsequent excursions.

8. Tournefortia argentea L. fil.

Littoral shrub, in the Dutch East Indies chiefly occurring on coralligenous beaches. The fruits [2]) possess floating power, hence can be spread by ocean-currents; Guppy found them in the drift-zone on the Keeling Islands. That this species is represented by robust specimens on the shores of Krakatao (and of Verlaten Eiland) is perhaps in some measure due to the rather high percentage of CaO (2,88 – 3,36 %) of part of the eruptive products [3]). Concerning its occurrence in the inner, higher parts of the island, as mentioned by Treub, I refer to what I have said on page 50 about *Wedelia*.

FERNS.

1. Gymnogramme calomelanos Kaulf.

At present named *Ceropteris calomelanos Und.* Heliophilous terrestrial fern originating from America, introduced in the Dutch Indies already long ago and at present naturalized throughout Java in not too dry regions from the plains up to 1000 m. above sea-level. It occurs in sunny or slightly shadowed, not too dry, by preference stony or little weathered localities; it is often plentifully found on perpendicular banks. The reproduction is effected exclusively by spores. In 1886 and much later yet it found on Krakatao many habitats quite fit for it; according to Raciborski [4]) it was in 1897 „sehr häufig „auf Krakatao und Verlaten Eiland." It is quite possible that in 1886 it was already frequent on Krakatao.

2. Acrostichum scandens J. Sm.

At the present time named *Stenochlaena palustris Bedd.* Terrestrial fern, by preference growing in shadowed, continually humid or swampy localities or at least in such spots where the water-table is near the surface. It occurs not only in the coastal plains, often not very far remote from the sea or even quite near it, but also much further in

---

1) Neue Flora Vulkaninsel Krakatau (1907), p. 44.
2) Described by Schimper, Indo-Malayische Strandflora (1891), p. 174.
3) Verbeek, Krakatau, Dutch Edition (1888), p. 292.
4) Die Pteridophyten der Flora von Buitenzorg (1898), pp. 69, 70, under the name of *Gymnogramme calomelanos*,

the interior, up to an altitude of 700 m, locally often in great numbers. In the surroundings of Batavia and Tandjong Priok it is rather common, especially along pools and ditches with sweet or slightly brackish water. Usually it climbs with the aid of its rhizomes high against the trees; where these are absent it can form – as was already mentioned thirty years ago by the excellent observer Raciborski [1]) – dense groups with creeping rhizomes. The reproduction is effected as well by these rhizomes as by spores which are easily spread by water. In May 1908 I found this plant in ravine-mouths on Lang Eiland, on which island it had been found already in 1896 by Boerlage, also in ravines. Probably the rather feeble, sterile specimen of Treub was gathered in a such-like habitat. According to Raciborski [1]) the plant was in 1897 numerous on the north-western side of Krakatao.

3. **Blechnum orientale** L.

Terrestrial fern, in Java one of the most common kremnophytes [2]) from the plains up to an altitude of 1600 m., except in regions where the east-monsoon is very strong. As a rule it grows on perpendicular banks and steep roadsides where it often locally predominates. The reproduction is effected exclusively by spores. In 1886 the north-western part of Krakatao offered it many localities fit to grow in and it is quite possible that it occurred then in many specimens.

4. **Acrostichum aureum** L. [3]).

Hygro-halophyte, chiefly growing in coastal marshes and also in salt or brackish, swampy localities in the interior [4]), up to many hundreds of metres above sea-level, rather rarely in sweet water and then only in scattered specimens. Along the clayey coasts of Java this very coarse fern is quite common in many localities, frequently rather far from the sea; the surface of the water is often coloured brown by myriads of floating spores, the only means of reproduction. In 1886 it may have grown in the north-western part of Krakatao in the mouths of the lower ravines with a saline soil (Cf. p. 50, footnote 3).

---

[1]) Die Pteridophyten der Flora von Buitenzorg (1898), p. 55, under the name of *Acrostichum scandens J. Sm.*

[2]) Cf. p. 31.

[3]) By some mistake Treub wrote *Acrostichum aureum Cav.* This fern, at present named *Ceterach aureum L. v. Buch*, is an inhabitant of the Canaries. The specimen collected by Treub was *Acrostichum aureum L.*

[4]) For such localities mentioned already almost a century ago by Blume [Flora Javae, Filices, p. 40 (under the name of *Acrostichum inaequale Willd*)].

5. **Pteris longifolia** L.

According to H i e r o n y m u s (Hedwigia LIV (1914) 283–294) all specimens from the Dutch East-Indies mentioned under this name belong to *Pt. vittata L., Pt. longifolia* being restricted to the New World. *Pteris vittata L.* is a kremnophyte spread throughout Java from the plains up to 2000 m., especially beneath 1200 m.; it reproduces itself exclusively by spores. It chiefly occurs on old brick walls, on cliffs, on steep roadsides and in stony or calcareous localities. R a c i b o r s k i [1]) calls it: „Eine sehr gemeine Art der Ebene „und des Hügellandes, seltener im Gebirge, doch noch bis 2000 m. „vorhanden. Besonders gerne an sonnigen Stellen, Kalkfelsen, Mauern; „im Gerölle am Gunung Guntur bis zum Krater aufsteigend". From all this it appears that in 1886 the fern found on Krakatao many fit habitats; it may have occurred in many specimens.

6. **Nephrolepis exaltata** Schott.

Terrestrial or less often epiphytical heliophilous fern, most variable as to its indumentum, producing many long stolons and consequently very often growing gregariously. It does not claim a fertile soil; it is often found growing in stony or rather sterile sunny, provided not too dry localilies. From the plains up to an altitude of ± 1200 m. it is one of the commonest ferns of Java. Besides by stolons it reproduces itself by spores. In 1886 the lower parts of Krakatao were very fit for it; it has been collected there at every excursion. The supposition of T r e u b [2]) that, like *Acrostichum aureum L.*, it prefers a saline substratum is quite unfounded; in such localities the plant never occurs, though it may be found in the interior of *old* coral islands.

7. **Nephrodium calcaratum** Hook.

This plant clearly illustrates what I have said on page 8 on the insufficient knowledge of the flora shown by different authors on the renewal of the flora of Krakatao. *Nephrodium calcaratum Hook.* (at present named Dryopteris calcarata O. K.) is a very shallowly rooting, heliophobous, hygrophilous fern, which has been collected in Java and Sumatra in several localities between 150 and 1000 m. above sea-level, exclusively on very humid river banks and on wet rocks in rivers, always in well-shaded localities. The most trustworthy

[1]) Die Pteridophyten der Flora von Buitenzorg (1898), pp. 151, 152 under the name of *Pteris longifolia L.*

[2]) Ann. Jard. Bot. Buitenzorg VII (1888), p. 219.

Raciborski [1]) gives as habitat „an feuchten, schattigen Schlucht- „wänden" and the no less reliable Posthumus [2]) says: „In primary „forest, on and between large rocks in the shadowed beds of small „rivers". Consequently it is very improbable that Treub should have found this fern in 1886 on the lower parts of Krakatao which were then everywhere very sunny and where consequently in the east-monsoon *the upmost layers* of the soil often became very dry or, at least, were not very humid. Never was this plant found back on Krakatao. When I examined the fern-materials collected in 1886 by Treub, it became evident that the constant appearance of this species in the list of plants found on Krakatao, even in those composed by authors [3]) who had continually free access to Treub's materials, is only due to incorrect reading. Burck, who determined Treub's ferns, did not name this plant *N. calcaratum*, but, and correctly, *N. cucullatum*, a name absent in the lists given by Treub. This last-named plant (at present named *Dryopteris unita* O. K.) is a heliophilous, mesophilous fern of often rather sterile localities. In Java it is „eine sehr gemeine Art der sonnigen Stellen der Ebene und „des Hügellandes, seltener in der unteren Gebirgszone, z. B. in dem „Gerölle des G. Guntur" [4]); in 1886 it found on Krakatao many ideal habitats and also in later years it was found there at every excursion.

8. Nephrodium flaccidum Hook.

This plant, at present named *Dryopteris flaccida* O. K., is a terrestrial fern of humid, well-shaded localities; in Java it has not yet been found beneath an altitude of 1000 m. Raciborski [5]), who is always very reliable, calls it „ein gewöhnlicher Erdfarn der unteren „und besonders der mittleren Gebirgszone, an schattigen Waldstellen „wild wachsend" and I also found it exclusively in such-like localities. Hence it is very improbable that Treub should have found it in 1886 on the lower, then still treeless and shadowless parts of Krakatao. In the list of Penzig [6]) it is mentioned, without any comment, as *not* having been found back.

[1]) Die Pteridophyten der Flora von Buitenzorg (1898), p. 170 under the name of *Aspidium calcaratum Bl.*

[2]) O. Posthumus, Betreffende de Pteridophyta van Djambi in Verslagen Kon. Akad. van Wetenschappen XXXVII, p. 81.

[3]) Docters van Leeuwen, Handelingen Eerste Ned.-Ind. Natuurw. Congres 1920, p. 75; Ann. Jard. Bot. Buitenzorg XXXI (1921) p. 239.

[4]) Raciborski. Die Pteridophyten der Flora von Buitenzorg (1898), p. 184 [under the name of *Nephrodium cucullatum (Bl) Baker*].

[5]) Die Pteridophyten der Flora von Buitenzorg (1898), p. 176 under the name of *Aspidium flaccidum Bl.*

[6]) Ann. Jard. Bot. Buitenzorg XVIII (1902), p. 108.

When I examined the specimen collected by T r e u b it appeared, as might be expected, that the determination was incorrect; the plant is a feeble specimen of *Dryopteris setigera O. K.*, a kremnophyte rather common in Java from the plains up to an altitude of 1800 m; its reproduction is exclusively effected by spores. This species was collected also at the excursion of 1897 and determined by R a c i b o r s k i as *Aspidium vile O. K.* (= Dryopteris vilis O. K., most probably only a xeromorphous form of *Dryopteris setigera*), under which name it is mentioned in the list of P e n z i g [1]) and by R a c i b o r s k i [2]). The well-known Buitenzorg fern-specialist V a n A l d e r w e r e l t v a n R o s e n b u r g h determined in 1919 the plants, which R a c i b o r s k i had found on Krakatao and Mount Guntur and which were recorded by P e n z i g and R a c i b o r s k i under the name of *Aspidium vile*, as *Dryopteris setigera*. R a c i b o r s k i calls his plant: „eine sehr interessante xerophile [3]) Art, im Gerölle „am G. Guntur ganze Strecken ..... bedeckend. Sehr reichlich am „Krakatao wachsend" and K o o r d e r s [4]) says of a specimen collected by him on Mount Lamongan „Auf einem fast vegetationlosen, „etwa 4 Jahr alten, sehr trocknen, felsigen Lava-Rücken". I also found this fern only on steep, rather sterile banks and in localities oecologically similar to these.

For typical kremnophytes as *Dryopters setigera* Krakatao was in 1886, and much later yet, an ideal habitat; it was found back there also in 1908 and 1919.

9. Pteris aquilina L.

At the present time named *Pteridium aquilinum Kuhn*. This is a halophobous fern, growing (often gregariously) in sunny or slightly shadowed, lime-deficient, as a rule well-pervious and not very fertile localities; sometimes, but rarely, it occurs on steep banks. It propagates by ramification of the far creeping rhizomes and by spores

---

[1]) Ann. Jard. Bot. Buitenzorg XVIII (1902), p. 108.

[2]) Die Pteridophyten der Flora von Buitenzorg (1888), p. 173 under the name of *Aspidium vile*.

[3]) This statement is not quite correct. G. G u n t u r receives rather much rain throughout the year. The 4 driest east-monsoon months have together 20—40 rain-days with a mean fall per rain-day of 10—14 mm. The lava-stream on Mount Lamongan mentioned by K o o r d e r s has, on an average, 20 rain-days in the 4 driest months and a mean fall per rain-day of 15—16 m.M. Consequently the somewhat deeper parts of the soil and the crevices in the rocks (in which the fern grows) are never very dry and there is no reason to consider the vegetation as xerophilous. *One should not confound the dry surface of the rocks with the moist crevices*. The fern in question is rather a kremnophyte or a chasmophyte than a xerophyte.

[4]) Annotation in Herbarium Bogoriensi ad Kds 24103 β.

and is spread throughout Java between 50 and 3000 m. above sea-level. Elsewhere in the Archipelago it occurs also beneath 50 m. The only specimen collected by Treub on Krakatao is very poor and hardly 20 cm. long. Penzig records this species also for the trip of 1897 but in the Buitenzorg Herbarium there is no specimen collected in that year and Raciborski [1]) does not mention it for Krakatao. This, of course, by no means proves that it was not found in 1897, but we may safely infer that this most conspicuous and easily recognizable fern was not numerous in the localities then investigated. After 1897 *Pteridium* was never found back on Krakatao what may be sufficiently accounted for by the fact that this part of Krakatao was no more thoroughly investigated but perhaps may be also caused by the high CaO-percentage (2,88 – 3,36 %) of part of the eruptive products [2]). I do not think it probable that in 1886 this fern took a large part in the vegetation of the island; if such had been the case, it would also have been numerous in 1897, *when the character of the flora was practically the same as in 1886* [3]) and, if so, Raciborski would certainly have collected and mentioned it.

10. Pteris marginata Bory.

At the present time named *Pteris tripartita Sw.* This is a shallowly rooting halophobous terrestrial fern of sunny or slightly shadowed humid localities, spread throughout Java from the plains up to an altitude of ± 1200 m. The reproduction is effected exclusively by spores. Raciborski [4]) calls this plant: „Ein häufiger Erdfarn an „feuchten, nicht sehr schattigen Stellen der Ebene". For such-like plants the soil of the lower parts of Krakatao was in 1886, generally speaking, rather unfit; the specimen, collected by Treub at Zwarte Hoek (as appears from the accompanying label) was still young, perhaps it was an unicum. If large specimens had been present, the very conscientious Treub would have collected them. In 1897 this species was not found back; hence it is improbable that in 1886 it took a large part in the vegetation, because the fern-vegetation in 1897 had practically the same character as during the expedition of Treub [3]). Neither was *Pteris tripartita* afterwards found on Krakatao.

[1]) Die Pteridophyten der Flora von Buitenzorg (1898), pp. 156, 157 (under the name of *Pteris aquilina L.*

[2]) Verbeek, Krakatao (1888) p. 292. To local presence of much CaO in the eruptive products points also the occurrence in 1906 of the moss *Bryum coronatum Schwaegr.* Cf. Chapter VIII.

[3]) Cf. p. 32, footnote 2.

[4]) Die Pteridophyten der Flora von Buitenzorg (1898), p. 158 under the name of *Pteris marginata Bor.*

11. Onychium auratum Kaulf.

At present named *O. siliculosum C. Chr.* This is a rather rare kremnophytic fern, occurring throughout Java from the plains up to an altitude of 2100 m. It grows almost exclusively on steep banks and is reproduced only by spores. Krakatao offered it in 1886 on its steep ravine-walls many fit habitats and according to Raciborski [1]) it was in 1897 very frequent on that island.

Of lower Cryptogams Treub found 2 mosses which he sent, for determination, to Dr. Van de Sande Lacoste in Amsterdam. But as this savant died shortly afterwards, these mosses have at present, to my knowledge not yet been named and, most probably, are lost. Besides these mosses Treub found six *Schizophyceae* (belonging to 3 genera) which formed in the localities investigated almost everywhere a thin greenish or glaucous gelatinous hygroscopical layer. These algae were provisionnally named by Treub: *Lyngbya intermedia Treub, L. minutissima Treub, L. Verbeekiana Treub, Symploca* spec. [2]), *Anabaena* spec. and *Tolypothrix* spec. Of the 3 first of these species Treub gave a short diagnosis. Afterwards these algae were, as it seems, described by Treub in Annals and Magazine of Natural History (1898), p. 148, a work I have not been able to consult. In volume V (elaborated by Forti, 1907) of the Sylloge Algarum of De Toni 2 of Treub's algae are renamed. *Lyngbya minutissima Treub* is reduced there (p. 335) to *Hypheothrix litoralis Hansg.* and *Lyngbya intermedia Treub* (p. 275) to *Lyngbya subolivacea Hansg. Lyngbya Verbeekiana Treub* I did not find mentioned in De Toni. Especially *Lyngbya Verbeekiana Treub* and *Hypheothrix litoralis Hansg.* were in 1886 very common in the investigated part of Krakatao.

Treub expressly states that in the entire island he did not find a single lichen. But as he examined only a small part of the island we may by no means conclude from his statement that in 1886 lichens did not yet occur on Krakatao.

It seems that Treub did not make researches after the micro-flora of the soil and after the fauna, at least he is silent about these.

Beneath I give a synopsis of the 34 plants found by Treub. Only the 11 Polypodiaceae (No. 9—19) are present in the Buitenzorg Herbarium. Whether of the other 23 plants materials are still extant is unknown to me.

---

[1]) Die Pteridophyten der Flora von Buitenzorg (1898), p. 164 under the name of *O. auratum Klfs.*

[2]) By a typographical error the name is in Treub's paper misspelled *Symploea.*

Oscillatoriaceae.

1. *Lyngbya subolivacea Hansg.*
2. *Lyngbya Verbeekiana Treub.*
3. *Hypheothrix litoralis Hansg.*
4. *Symploca* spec.

Nostocaceae.

5. *Anabaena* spec.

Scytonemaceae.

6. *Tolypothrix* spec.

(Nos. 1–6:) In the interior, forming a slimy layer on ashes and pumice, especially No. 2 and 3 very common.

Bryophyta.

7. 8. Two undetermined and probably lost mosses, which, according to Treub himself [1]), took but an unimportant part in the new vegetation.

Polypodiaceae.

9. *Dryopteris setigera O. K.*
   Interior. See p. 57, under No. 8.
10. *Dryopteris unita O. K.*
   Interior. See p. 56, under No. 7.
11. *Nephrolepis exaltata Schott.*
   Interior. See p. 56, No. 6.
12. *Blechnum orientale L.*
   Interior. See p. 55, No. 3.
13. *Stenochlaena palustris Bedd.*
   Interior. See p. 54, No. 2.
14. *Ceropteris calomelanos Und.*
   Interior. See p. 54, No. 1.
15. *Onychium siliculosum O. K.*
   Interior. See p. 60, No. 11.
16. *Pteris tripartita Sw.*
   At Zwarte Hoek. See p. 59, No. 10.
17. *Pteris vittata L.*
   Interior. See p. 56, No. 5.
18. *Pteridium aquilinum Kuhn.*
   Interior. See p. 58, No. 9.
19. *Acrostichum aureum L.*
   Interior. See p. 55, No. 4.

---

[1]) Ann. Jard. Bot. Buitenzorg, 1me Série VII (1888), p. 219, footnote 1.

Gramineae.

20. *Pennisetum macrostachyum Brongn.*
Beach and interior. See p. 48, No. 8, and p. 53, No. 6.

21. *Neyraudia madagascariensis Hook, fil.* var. *Zollingeri Hook. fil.*
Interior. See p. 53, No. 7.

Cyperaceae.

22, 23, Two unnamed species.
Beach. See p. 47, No. 5 and 6.

Hernandiacae.

24. *Hernandia peltata Meissn.*
Beach. See p. 47, No. 4.

Leguminosae.

25. *Erythrina* sp.
Beach. See p. 46, No. 1.

Guttiferae.

26. *Calophyllum inophyllum L.*
Beach. See p. 44, No. 6 and p. 46, No. 2.

Apocynaceae

27. *Cerbera manghas L.*
Beach. See p. 46, No. 3.

Convolvulaceae.

28. *Ipomoea pes caprae Sw.*
Beach. See p. 48, No. 7.

Borraginaceae.

29. *Tournefortia argentea L. fil.*
Interior. See p. 54, No. 8.

Goodeniaceae.

30. *Scaevola frutescens Krause.*
Beach and interior. See p. 49, No. 9 and p. 53, No. 5.

Compositae.

31. *Wedelia biflora D. C.*
Interior. See p. 49, No. 1.

32, 33, 34. Three insufficiently known species of which 2 probably belonged to the genera *Blumea, Erigeron* or *Pluchea,* the third to *Senecio* or *Erechthites.*
Interior. See p. 50, No. 2 and 3 and p. 52, No. 4.

From his finds Treub drew a series of conclusions numbered by me from 1 to 9. On each of these I state my views herebeneath.

1. The flora of the interior of Krakatao entirely differs from that of the beach.

Apparently this fact is emphasized by Treub because he considers it as contrary to his original opinion [1]) that the new vegetation of Krakatao, like that of a very young coral island, would originate from seeds washed ashore and would consist mainly of littoral plants. Many such plants, as Treub could have observed in the Buitenzorg Botanical Gardens, may prosper — at least when cultivated — far from the sea and at a rather considerable altitude [2]). Hence it might be expected that on Krakatao they would be able to penetrate to a rather large distance from the sea. But there is a great difference between Krakatao where, in consequence of leaching caused by the large rainfall, the higher layers of the soil in the inclined localities soon became salt-free and a very young and small level coral island which consists, so to say, entirely of a young beach, hence contains much salt, does not possess in its centre a basin of sweet water (stored rainwater) and therefore can bear only a halophilous vegetation. On the beach of coral islands as well as on that of Krakatao seeds and fruits of all kinds of plants may be washed ashore and the lighter ones of these, after having dried during the low tide, may by the sea-wind be driven into the interior. But in the interior of a very young coral island the seedlings of halophytes only can remain in life; the other ones soon perish in consequence of the large salt-percentage of the soil. In the interior of Krakatao the seedlings of non-halophytes also can develop. — The occurrence of halophytes in the interior depends on many conditions, apparently not only on competition with other plants or the mode of dispersal of the seeds. In Java many littoral plants (i.a. *Albizzia retusa Bth.*, — *Euphorbia Atoto Forst.*, — *Guettarda speciosa L.*, — *Pemphis acidula Forst.*, — *Pongamia pinnata Merr.*, — *Scaevola frutescens Krause*, — *Sophora tomentosa L.*, — *Spinifex littoreus Merr.*, — *Thespesia populnea Soland.* and *Tribulus cistoides L.*) are in a wild state never found at a distance worth mentioning from the sea, not even when the locality behind offers plenty of space; neither are these species ever found in saline localities in the interior where some other halophytes of the coast (*Acrostichum*

[1]) Ann. Jard. Bot. Buitenzorg, 1me Série VII (1888), 216.
[2]) Buitenzorg lies at about 250 m. above sea-level.

aureum L., – *Cynanchum carnosum Schlecht.,* – *Diplachne polystachya Backer,* – *Gymnanthera paludosa Schum.,* – *Paspalum vaginatum Swartz,* – *Scirpus littoralis Schrad.,* – *Xerochloa imberbis R.Br.*) may thrive. Besides littoral species and such halophytes as occur also in the interior, one finds on broad beaches, especially on their older less saline or quite leached-out parts, a considerably large number of plants which may also be found in the interior in not salt localities. Treub, who examined only a small part [1]) of the then still very young beach of Krakatao, found there already such a plant of the interior (*Pennisetum*), whilst, on the contrary, he mentions some typical plants of the coast (*Tournefortia argentea L. fil.* and *Wedelia biflora D.C.*) for the interior only („la montagne proprement dite"). This does not warrant the conclusion that Treub by the plants found by him has proved the incorrectness of his statement, for he was no florist and may have misremembered. Moreover it is not certain that he, who was no phyto-geographer, has taken the word beach in its oecological sense. It often occurs in the tropics that localities, which verge on the beach but which by some cause f.i. by a higher situation or a too great age, considerably differ from it, yet are considered as such by inexperienced collectors and Treub may have made the same mistake. After what I have remarked on pp. 5 and 6 about young vegetations in the tropics it need scarcely be said that on the eastern beach of Krakatao, which was not examined by Treub, many species may have grown which were not found by him at Zwarte Hoek. The same holds good for the interior of Krakatao.

2. *It is beyond doubt that the fruits of the 4 Compositae have been carried over to Krakatao by wind.*

This conclusion is apparently not based on observations or on experiments but is a more supposition. Hence it cannot furnish proof for any hypothesis but only increases the number of unproved guesses. The first of the Compositae found by Treub, *Wedelia biflora D. C.,* is a common littoral plant, occurring also along rivermouths. As I have already said on page 50, it possesses rather large fruits without a pappus but with a thick pericarp developed into a floating tissue, a fact which must have been unknown to Treub. This species was certainly not introduced by wind but most probadly by the sea.

[1]) On the western and the southern side of Krakatao the coasts of the isl'and are very steep, on the south-eastern and eastern side a well-developed beach is present over a length of abouth $3^3/_4$ km. The beach of Zwarte Hoek is $\pm$ 750 m. long. This latter beach only was examined by Treub, consequently not yet 17 % of the entire beach. Of the interior he examined only a very small part.

The identity of the 3 other Compositae is not established but it is certain that they all possess a pappus on the fruits. From this, however, by no means follows that these fruits had been carried by wind across the sea for length of at least 20 km., the distance by which Krakatao is separated from the island Sebesi, the nearest point of which it is sure that the vegetation in 1883 was not entirely destroyed. As I have set forth on pp. 50–52 Treub must have found 3 out of the there-mentioned 5 species. Three of these five (*Blumea balsamifera D. C.,* – *Erigeron linifolius Willd.,* – *Pluchea indica Less.*), perhaps also the 2 other ones, may quite well have grown before the eruption in or near the locality investigated; their fruits and even the subterranean parts of some of them may, in well-sheltered spots, have survived the catastrophe. If one will insist on introduction having taken place from outside, one should bear in mind that these 5 Compositae are quite common in West-Java and may occur also in South Sumatra. Their fruits may very well have been carried by flowing rain-water to the rivers and by these to the sea; they may also have been driven by wind straight into the sea. The sea may have washed them ashore at Zwarte Hoek and, after they had become dry, the wind may have driven them into the interior [1]. Ernst has argued on very good grounds [2] that transport by water from Java to Krakatao can be made within a single day and it has never been proven that the fruits of the Compositae mentioned have lost their germinative power after having floated for a day on sea-water. – Moreover the possibility exists that the many visitors who came to Krakatao after the eruption but before Treub's investigations have carried over fruits on their clothes or with their things. Finally, fruits may have been introduced by native vessels landing at Krakatao, by driftwood, by pumice, by birds. *There does not exist the faintest shadow of a proof for direct introduction by wind.* Hence I can by no means share the opinion of Schimper [3] and Rikli [4] that the possibility of direct transport of seeds and fruits by wind for a long distance across the sea is *„endgültig nach-„gewiesen"* by Treub's investigations. These investigations or, to speak the plain truth, *guesses,* lack all conclusive force.

---

1) The beach at Zwarte Hoek is for the greater part secluded from the interior by a steep wall. In the most north-western part however the lower basaltic slopes are easily accessible from the beach.

2) Neue Flora Vulkaninsel Krakatau (1907), p. 56.

3) Schimper, Pflanzengeographie (1908), p. 90.

4) Rikli, Richtlinien der Pflanzengeographie in Fortschr. Naturw. Forschung III (1912), p. 260.

The introduction of Compositae with a pappus on the fruits is by many authors on the new flora of Krakatao ascribed to direct action of the wind and sometimes they add as a corroborating argument the unlimited numbers Nature works with. One can, of course, to a certain extent acknowledge the possibility of such a transport, but when, as in the case of Krakatao, *observations as well as experiments are entirely lacking,* oecological and phytogeographical problems cannot be solved by the acknowledgment and discussion of probabilities. Not before it has been shown by irreproachable experiments that seeds, fruits and fern-spores *are* directly transported by wind to Krakatao from the surrounding islands or at least, that, *in such circumstances as prevail in the Sunda-straits,* they *might* be transported by wind that distance *across the sea,* one has some right to take for granted that such a transport has really taken place.

3. The most interesting fact is the occurrence of 11 ferns. Only 2 of these (Acrostichum aureum L. and Nephrolepis exaltata Schott) are widely dispersed in the islands of the hot regions[1]*. Only for these 2 which probably prefer a more or less saline substratum Krakatao can have offered some advantages in this respect. For the 9 other species this is not the case.*

This is a rather vague conclusion. *It is not true* that but 2 of the 11 fern-species found by Treub are widely dispersed in the islands of the hot regions; *all of them,* without exception, *are quite common in many localities of Java and Sumatra.* One should bear in mind that when Treub in 1888 wrote down this statement, the flora of the East Indian Archipelago and even that of Java was but very insufficiently known, especially as regards the distribution of the species. The older collectors in the Dutch Indies, Junghuhn excepted, have applied themselves much more to discover new or rare species than to get a survey of the local composition of the vegetation.

Of the 2 supposed halophilous ferns mentioned by Treub only *Acrostichum aureum L.* prefers a saline substratum, *Nephrolepis exaltata Schott* is halophobous. Seven of the 11 ferns found in Krakatao in 1886 (No. *9, 10, 11, 12, 14, 15* and 17 of the list on pp. 61, 62) are in Java and Sumatra common plants of unfertile soils. To these ferns the basalt, pumice and young unweathered ashes of the investigated part of Krakatao offered very favorable conditions of growth.

[1]) Ernst [Neue Flora Vulkaninsel Krakatau (1907), p. 7] somewhat inadequately translates: „die mit 11 verschiedenen, im indomalayischen Gebiet weit verbreiteten „Arten vertreten waren, unter denen aber nur zwei in der *Strandflora* der Inseln vor„zukommen pflegen". This Treub did neither say nor mean.

No. 13 and 19 of Treub's list, *Stenochlaena palustris Bedd.* and *Acrostichum aureum L.*, ferns of humid or swampy localities, could find suitable habitats in humid parts of the lower ravines, the last-named fern especially in such places where, after the eruption, NaCl was deposited by water oozing out (see p. 50, footnote 3).

Of No. 18, the calcifugous *Pteridium aquilinum Kuhn*, the very poor specimen collected by Treub bore in 1924 on its rhizome still vestiges of grey volcanic ashes. If these did not adhere to it by some accident one would conclude from their presence that Treub collected the plant on a field of ashes. The young ashes of Krakatao offered to this species probably a rather unfit substratum because of their high CaO-percentage.

For No. 16, *Pteris tripartita Sw.*, a halophobous, non-kremnophytic, shallowly rooting fern of sunny or slightly shadowed, constantly humid localities, the lower parts of Krakatao were in 1886, as a whole, rather unfit; the pumice-covered ridges because of the very porous upper layers of the soil becoming too dry after a few rainless days, the bottom of the ravines because of the local deposits of NaCl. But, in some places at least the locality must have been favorable to its growth, the young specimen collected by Treub being rather robust. One should bear in mind that Treub made his collection in June or July, shortly after the end of the rainy monsoon.

Summarizing, the investigated part of the island offered to 9 out of the 11 species in many localities favorable conditions of growth; hence these 9 species may have occurred in great numbers. All of them were found back in 1897 by Penzig or Raciborski.

Treub takes for granted that the spores of the ferns found by him were carried over from the surrounding islands to Krakatao by wind. This he does not wonder at, for he believes that the much heavier seeds or fruits of the Phanerogams were introduced in the same manner. I argued already (p. 65) that this last mentioned supposition is not in the least proven and though I fully admit for very light and fine fern-spores the possibility of transport by wind over rather large distances over land, I emphasize that in the case of Krakatao it is not proven that direct transport of spores by wind from outside the island, i.e. over a distance of at least 20 km. across the sea, has really taken place.

Without taking the wind, the water (p. 65) or man (pp. 10 and 33) for the agent of introduction, the occurrence of many species of ferns in Krakatao may be easily explained. No. 9, 10, 11, 12, 14, 15 and 17 of Treub's list, all of them typical kremnophytes, may quite well have been growing before the eruption in the locality investigated in 1886, an old basaltic slope, covered by the eruption with ashes and pumice; they may (as was already argued on pp. 31, 32) have survived the catastrophe in the form of rhizomes or of spores. Very soon after the eruption the covering layer was here and there washed away. In such spots the surviving rhizomes or spores may have developed into plants, the offspring of which may have spread over the permanently covered parts. I think it much more probable that the *rich* fern-vegetation which Treub found *within three years* after the eruption in the locality investigated has arisen in this manner, than that the spores of all these species were carried over by wind over a distance of at least 20 km. to suitable habitats in such numbers as to allow the vegetation in question to develop so rapidly quite anew. Alas nothing is known of the time the spores of these ferns need to develop into adult plants under the conditions of growth prevailing at Krakatao. On this point also the investigations after the re-establishment of the vegetation have been insufficient.

No. 16 and 18 *(Pteris tripartita Sw.* and *Pteridium aquilinum Kuhn)* may quite well be remnants of the old vegetation of the higher parts of Krakatao. In many primeval forests in the mountains of Java sunny or slightly shadowed spots occur, caused by large trees having fallen to the ground. Not rarely also, especially in swampy localities [1]) or on steep ravine-walls or dry ridges, the soil is locally unfit for a dense vegetation of trees. In such sunny spots it often bears a grass-, fern- or bush-jungle. In the former woods of Krakatao, which were never investigated and *of which we know absolutely nothing,* many such open or slightly shadowed spots may have been present and it is quite possible that they bore one of these 2 ferns (which never are found in each other's company: *Pteris tripartita* needing much more water than *Pteridium aquilinum*). Both ferns may have survived the eruption by means of their rhizomes or their spores; both of them may, after the eruption, have found fit habitats in the higher not saline parts of the island; the spores of both may have been

[1]) Such localities, by constant marshiness unfit for a dense vegetation of trees are found in Java i.a. on Mount Gĕdè, Mount Patuha, at the base of the twin-mountains Wajang-Windu and at the southern slope of Mount Hjang. It is remarkable that whilst sweet water marsh-forest has developed in many swampy parts of the *plains* (i.a. near Djatiroto in Java and in vast localities in Sumatra) such has not happened, to my knowledge at least, anywhere in the *mountains* of Java.

carried downwards by rain-water or by wind, till they reached the locality examined by T r e u b. Hence for these species too one is not bound to admit direct transport by wind over a large distance. But here also nothing is proven.

Of No. 13 and 19 of T r e u b's list, *Stenochlaena palustris Bedd.* and *Acrostichum aureum L.* ferns of humid or swampy localities, of water-sides or a soil with a high water table, the spores possess floating power, hence may be transported by water. The spores of these 2 ferns may quite well have been carried to Krakatao by sea-currents and it is not proven that by floating on the sea during one day they lose their germinative power. Neither is transport by animals, f. i. birds, quite impossible; S c h i m p e r already pointed out the possibility of this means of conveyance [1]).

I am fully aware not to have adduced conclusive proofs for the correctness of these views. But transport of spores by wind from outside of the island is just as little proven and the supposition that part of the new fern-vegetation has arisen from, at least *might* have arisen from rests of the old one is, I think, supported by probability. T r e u b *having taken a priori the most important and most dubious point for granted,* has neglected to assiduously seek for rests of the old vegetation, that might have remained in life. At present, almost half a century after the eruption, it is impossible to decide whether the original vegetation was in 1883 entirely destroyed or only for a part. Consequently it is uncertain whether the new flora necessarily was introduced. Hence the Krakatao-problem cannot be distinctly posed and therefore not be solved either now or in the future, even when the most extensive and irreproachable investigations would be carried out.

4. The new flora of Krakatao consisted almost exclusively of ferns. Phanerogams occurred only here and there in scattered specimens on the beach and on the mountain.

Passing over the fact that the first part of this assertion by its defective wording is contradictory to that given under point 7 (p. 75), I emphasize that this statement may be applied only to the comparatively small and, locally at least, very unfertile part of Krakatao which was examined by T r e u b. Of the vegetation of the higher parts of the mountain in 1886 we *know nothing more than that Treub already saw rather large plants growing there.* There is not a single

[1]) S c h i m p e r, Indo-Malayische Strandflora (1881), p. 157, under the name of *Chrysodium aureum Mett.*

proof that the vegetation of these higher parts also consisted almost exclusively of ferns; on the contrary it is possible that there from rests of the old vegetation a new flora had developed consisting for a large part of Phanerogams. Just as little is known of the vegetation of the lower parts of the south and east side of Krakatao; it is quite possible that a large number of higher plants were growing there. It should always be borne in mind that Treub examined only a very small part of Krakatao, which differed from the rest of the island by the presence of basaltic rocks soon laid bare. Probably nowhere else in the *lower* parts the covering layer had so soon disappeared. The permanently and deeply covered parts on the south and east side had to receive a new vegetation which could most easily come from the higher parts of those ravines from which the covering layer had been washed away very early (See pp. 20, 21). It is unknown which plants were growing in 1886 in these ravines and it is just als little known which plants occurred on the southern and eastern lower part of Krakatao. They may have been Cryptogams, they may have been Phanerogams, probably both. Nobody knows anything about it!

For sure we only know that Treub in the restricted locality examined by him found a vegetation of kremnophytes, mainly ferns. Not Penzig in 1897, as Docters van Leeuwen supposes [1]), but Treub in 1886 was the first who found in Krakatao a plant-association or, more exactly speaking, already three; one association of algae and two of higher plants: on the beach a very poor one, for the main part composed of introduced halophilous Phanerogams; in the interior a richer one, mostly of heliophilous kremnophytic ferns, perhaps originating for the greater part from surviving components of the old vegetation.

That in the investigated locality Phanerogams occurred only in seattered specimens between the ferns is quite naturally explained by the fact that in these parts of the Dutch Indies there is a much higher percentage of kremnophytes among the ferns than among the higher plants. Of the 42 kremnophytes of the lower regions recorded on p. 31 as many as 29 are ferns or fern-allies. *Moreover, under circumstances, ferns can, by their spores, reproduce themselves easier and sooner than Phanerogams which often need the help of animals for the pollination of their flowers or for the dispersal of their fruits or seeds.* The fauna of Krakatao in 1886 being quite

[1]) Handelingen Eerste Nederlandsch-Indisch Natuurwetenschappelijk Congres (1920), p. 45. – Annales Jard. Bot. Buitenzorg XXXI (1921), p. 115.

unknown, it will be fruitless to seek for the relations existing then between that fauna and the flora. With some right one might suppose that the number of pollen-transporting and seed-dispersing animals was then very small, but suppositions are no proofs. Moreover, but few *reliable* observations have been made in the Dutch East Indies on the biological relations between flowers and fruits on the one side, animals on the other side. In the publications of Knuth [1]) and Docters van Leeuwen [2]) on this subject a number of flower-visiting insects have been recorded but often these authors give no proof to have ascertained that the insects, when visiting a flower, pollinated the stigma and never they seem to have examined *whether the formation of fruits really depended on such pollination.* In a recent paper on endozoic seed dispersal [3]) Docters van Leeuwen gives a list of fruits and seeds found by him in the intestines of some birds shot, from which he deduces that these birds disseminate the plants. But fruit-eating birds need not necessarily be seed-dispersing birds [4]). It is clear that, generally speaking, only a very careful sowing of seeds found in bird-droppings can procure the required evidence for the possibility of dispersal of a given plant by birds.

It is quite possible that in other, more fertile parts of Krakatao Phanerogams predominated or, at least, there is no proof that the vegetation there also was mainly composed of ferns. The statement of Docters van Leeuwen [5]): „In 1886 Krakatao was furn- „ished . . . . . in the interior with algae and ferns. A few grasses and „bushes were also to be met with already here and there" does not hold good for the entire island but only for the small part investigated by Treub. Nothing is known of the vegetation of the higher ravines which were the most important localities for the investigation.

5. Hence, ferns have the task to prepare the soil for the new vegetation of Phanerogams, which afterwards will cover the island like it did before the eruption.

This conclusion seems to be very badly founded. Treub apparently supposes that, whilst the flora of the investigated part of

1) Knuth, Handbuch Blütenbiologie III, 1, 2, passim.

2) Docters van Leeuwen, Blumen und Insekten auf einer kleinen Koralleninsel, Ann. Jard. Bot. Buitenzorg, XXXVII (1927), p. 1, seq.

3) Docters van Leeuwen, Kleiner Beitrag zur Kenntnis der endozoischen Verbreitung einiger Hochgebirgspflanzen auf Java [Festschrift zum siebzigsten Geburtstag von Karl von Goebel (1925), p. 82].

4) On this point one should consult Hans Winterstein, Handbuch der vergleichenden Physiologie II (1911), p. 1198 seq. „Die mechanische Funktion des Muskelmagens der Vögel".

5) Ann. Jard. Bot. Buitenzorg XXXI (1921), p. 115,

Krakatao in 1886 consisted mainly of ferns, the island was before the eruption totally covered with a Phanerogamic vegetation. But this is not all proven for the basalts of Zwarte Hoek, nor is it, considering the vegetation in such-like localities in the same kind of climate, probable. And it is just as little probable that ferns *prepare* the locality for Phanerogams. By the mechanical and chemical action of the roots and also by retaining decaying vegetable matter and fruits or seeds which otherwise might be washed away, a kremnophytic vegetation can, of course, prepare the way for a subsequent vegetation which puts higher claims on the fertility of the soil. But this is only a mechanical-chemical-oecological succession, not a transition from ferns to Phanerogams. In Krakatao the kremnophytic vegetation could for the greater part consist of ferns because of the island receiving in the dry monsoon comparatively much rain; in other parts of the Dutch Indies (the much drier island Madura [1]) f.i.) the kremnophytic vegetation consists mainly of other plants. In Krakatao *there has not been a succession Cryptogams-Phanerogams as such, but only a survival and victory of such species present as were momentarily the most fit to live under the existing conditions.* As long as the soil is but little weathered and unfertile it bears a kremnophytic flora, which as the fertility increases, is sooner or later superseded by other, more exacting plants. The speed of succession depends mainly on the conditions offered by the soil; the sooner these are modified, the sooner the vegetation changes. The lava-streams which Verbeek in 1880 [2]) saw in the no more existing northern part of Krakatao must have been at least 200 years old and yet they bore only a very scanty vegetation. Of the rocks visited by Treub in 1886 Penzig said 11 years afterwards that the aspect of the vegetation for the greater part was unchanged. The general aspect of the vegetation of steep road- and riversides in the surroundings of Buitenzorg has, in the course of at least 20 years, not perceptibly changed. Rather more than a century ago Daendels built his renowned road from West- to East-Java. Wherever this road has been cut through a hill-side the *vertical* parts of its banks, if not bare, are still covered with kremnophytes. In localities more favourable to plant-growth the vegetation may doubtless change much more rapidly, but nowhere Ferns, *as such,* precede the Phanerogams.

It is not at all proven that the present forest and grass-jungle on

---

[1]) Madura is a comparatively small island lying north of the eastern part of Java.
[2]) See p. 16, footnote 3.

the east and south side of Krakatao were *preceded* by a vegetation of ferns. Higher plants may quite well have been among the first to appear. Nothing thereabout is known.

6. *It is not at all strange that numerous fern-spores drop on Krakatao as even the seeds of Phanerogams are carried over by wind. But it is almost inconceivable how ferns and their prothallia can live in the very arid soil of Krakatao which is almost continually heated by the glowing sun, whilst there is not the least shadow. Treub and Burck have in the laboratory at Buitenzorg sowed spores from Krakatao on many different substrata in order to ascertain whether the prothallia and seedlings of these ferns possess special adaptations enabling them to live in such unfavourable circumstances as presented by Krakatao. No such adaptations could be found.*

My views on the first part of this conclusion have already been stated herebefore (pp. 64, 65). – The basaltic rocks investigated by Treub were doubtless very unfertile. But, as Ernst[1]) rightly observed, the pumice- and ash-fields were by no means so poor as Treub thought. The ashes contained in soluble compounds nearly all salts indispensible for plant-growth and in sufficient quantities, barring nitrogen- and phosphor-compounds. Nitrogen-compounds could be supplied by water, either falling as rain which in the tropics often carries down much dust which was floating in the air, or flowing down the ravines during heavy showers. *Such higher parts of the ravines as were already laid bare or almost so must have contained many easily removable organic or anorganic rests of the former vegetation;* water containing nitrogen-compounds could reach the upper layers of the soil by capillary motion. To my knowledge no analysis has been made of the ground-water of Krakatao. But a well, sunk in 1896 on Lang Eiland, which island was in much the same circumstances as the lower ash- and pumice-fields of Krakatao, *yielded water undrinkable on account of its high percentage of organic substances* [2]). Ernst, moreover, draws attention to the animal and vegetable rests which were washed ashore, part of which might be driven into the interior by wind. Treub did not institute an investigation after the presence of soil bacteria, fixing free nitrogen. Hence one cannot ascertain to which degree by such bacteria the growth of higher plants in 1886 was promoted or made possible. But it is very probable that the nitrogen

---

1) Ernst, Neue Flora Vulkaninsel Krakatau (1907), p. 51.
2) Tijdschr. Koninkl. Ned. Aardr. Genootschap XIV (1897), p. 120.

organisms still remained in the soil. The sole effect of the eruption and the heating of the soil should be a partial sterilisation, the effect of which is an increased activity of the nitrogen-fixing and nitrification organisms.

The glowing heat on Krakatao is rather exaggerated by Treub, who, living usually an indoor-life, was no trained mountaineer and did not possess great physical force. He visited Krakatao in June or July, in the beginning of the east-monsoon, apparently on a sunny day, and must – as all untrained walkers in the tropics – when ascending the shadowless steep slopes have suffered much from fatigue and thirst. Hence he got the erroneous impression of the locality being *continually* heated by a burning sun. But in the west-monsoon the sky is often cloudy for days, even for weeks, and there falls very much rain. In that season there is no question of the soil drying out, the air is very humid and the fern-spores can easily germinate. Dew also brings water and, by its evaporation, coolness. When the roots of the young plants have grown to a certain length they can absorb sufficient water from the soil, only the upper layer of which dries much out in the east-monsoon. This monsoon is, moreover, frequently interrupted by rain-days. In the tropics many plants may be seen thriving on a hot and apparently very dry soil. On the glowing sand of the dunes of Madura [1]) I found in July 1916 after a drought-period of many weeks, several quite fresh seedlings of *Tribulus cistoides L.* On rocks and sand it may be frequently observed that plants can stand very high soil-temperatures; besides, only the upper layers of the soil are so strongly heated. Not the *heat,* but the *drought* caused by long rainless periods finally kills the plants. Man, always inclined to an anthropocentrical trend of thought and often insufficiently trained in bodily exercise, often measures heat by the trouble it causes him. Fatigue and thirst are apt to confuse the observations frequently made without instruments. A really *very high* soil-temperature is in West-Java a very rare phenomenon. In one of the islands of the Krakatao-group, Lang Eiland, the temperature, from which the observer [2]) suffered much, amounted in a sheltered locally only to 34° C and fell during the night to 30° C; the temperature of the soil heated by the direct action of the sun rose in the afternoon but little above 60° C. For tropical and subtropical plants this is by no means an extremely high temperature. In British

---

1) Cf. p. 72, footnote 1.
2) Tijdschrift Kon. Nederl. Aardrijksk. Genootschap XIV (1897), p. 120.

India air-temperatures of 50 – 52° C have been recorded. Massart [1]) saw in 1898 in the Sahara a rather rich vegetation of higher plants on a soil heated to 67° C. Peschuel Loesche found on the shore of Loango (West-Africa, ± 5° southern latitude) a profusely flowering *Ipomoea* on a sandy soil which, at its surface, had a temperature of 69° C. Several lichens thrive on rocks which are regularly heated up to 60° C, one species even on rocks heated up to 70° C [2]). Ernst [3]), it is true, says of Krakatao: „Der Boden wird jeden Tag, „ebenso auf hellen Tage der Regenzeit, auf Temperaturen erhitzt, „welche das Wärme-maximum für Keimung und Wachstum der meisten „Pflanzen bedeutend übersteigen," but this assertion is not corroborated by any experiment made by him in order to ascertain at which maximum-temperature tropical plants can germinate, live and grow under such circumstances as prevail on Krakatao. On this important point nothing has been examined, much less proven.

7. The ashes and pumice of Krakatao are almost everywhere covered with a thin gelatinous hygroscopical layer of greenish or glaucous algae (*Cyanophyceae*). *Treub does not doubt that only the presence of this layer enables the spores of the ferns to germinate. Hence the algae prepare the soil for the ferns about in the same manner as these, in their turn, will do for the Phanerogams.*

This discovery of Treub is of real importance. Properly speaking it is the only important result the botanical investigations of Krakatao have yielded up to now. Probably, *Cyanophyceae*, which need only a minimal quantity of soil-nitrogen, were spread throughout the island. In September 1896 Burck, Boerlage and Mulder found these algae also in another island of the Krakatao-group. Lang Eiland. There they formed a thin layer on the ravine-walls shielding these from being washed off by rainwater flowing dow (Cf. Chapter V) Half a year afterwards Penzig [4]) and his companions saw such a layer of algae in many spots in Krakatao; Ernst [5]) found it back in 1906.

Treub's supposition that the hygroscopical slimy layer of algae strongly contributed to enable the spores of ferns and mosses to germinate is probably correct. Treub found in a sample taken on

---

1) Voyage Botanique au Sahara in Bull. Soc. Royale Bot. Belgique XXXVII (1898) p. 285.
2) Schimper, Pflanzengeographie, 2e Auflage (1908), p. 49.
3) Neue Flora Vulkaninsel Krakatau (1907), p. 50.
4) Ann. Jard. Bot. Buitenzorg XVIII (1902), p. 101.
5) Neue Flora Vulkaninsel Krakatau (1907), p. 32.

Krakatao one [1]) germinating spore of *Pteris vittata* [2]). The layer of algae being permanently moist or at least drying out less than the soil proper, spores may have germinated on it also in the east-monsoon which, as I have already stated above, is interrupted by frequent rains. But in all probability the presence of algae was not a *conditio sine qua non* for the development of a fern-vegetation. Everywhere in Java may be observed that fern-spores germinate quite well in humid substrata *not* covered by algae and this may very well have happened also in Krakatao, at least in the West-monsoon. Admitting that perhaps the layer of algae has contributed to a large extent to a rapid extension of the fern-flora I hold that it was not an *indispensable* preparer of the locality. But by its presence spores may have germinated throughout the year whilst otherwise perhaps they could only have done so in the West-monsoon. The part taken in the tropies by algae in facilitating the germination of fern-spores should be further investigated.

8. *What would happen if Krakatao was situated at a very great distance from every plant-covered land? The Phanerogams established on the beach would ascend the mountain to a certain height driving back before them the ferns. This movement would be arrested as soon as the conditions above the height reached were unfit for littoral plants. The seeds of other Phanerogams, fit to live at greater altitudes, would never or but very rarely reach the island, lost in the ocean. Therefore in the higher part of the island the ferns, perhaps accompanied by some Lycopods, would predominate for ever; the Phanerogams would be restricted to the lower parts and the beach.*

This supposed case is found realized in Juan Fernandez and Ascension. The small volcanic island Juan Fernandez is situated at ± 625 km. from the coast of Chili and Moseley says of the flora: „There „are upwards of 24 species of Ferns growing in this small island and in „any general view the Ferns form a large proportion of the main mass „of vegetation". The soil of Ascension, an entirely volcanic island of about the same height as Krakatao, is described as follows: „St. „Helena has been called a barren rock, but it is a paradise as compared „with Ascension which consists of a scorched mass of volcanic matter

1) This one spore apparently possessed the fission-power of a bacterium. In Warming's Pflanzengeographie (3e Aufl. 688) it has already multiplied to: „Farnsporen welche sich in Menge einfanden".

2) The tetraedal spores of this fern are by the very prominent ridges they are provided with easily distinguishable from the spores of the other ferns found by Treub in Krakatao.

„in part resembling bottle-glass and in part cokes and cinders". Hooker says of the flora: A small green peak, 2800 feet above the sea, monopolises all the vegetation, which consists of Purslane, a grass and an Euphorbia in the lower parts of the island, whilst the green peak is clothed with a carpet of ferns and here and there a shrub. Treub mentioned these facts in order to show that the situation found by him on Krakatao forms no exception.

This conclusion and the instances cited show that Treub did not sufficiently take into account the innumerable factors by which a vegetation is defined. It is here not the place to discuss the problem of the vegetation of these and other islands but it is by far not so simple as Treub supposed. Rainfall, evaporation, temperature, soil conditions, origin and history of the flora, influence of man and of the fauna have i.a. to be taken in account. Climatic conditions are almost neglected by Treub and, as regards the instances cited, also character and age of the vegetation. He reckons only with introduction from without, all other factors are dispensed with.

Apparently Treub was of opinion that the sea washed ashore almost exclusively seeds of littoral plants. Everyone who takes the trouble to minutely examine the drift-zone will soon arrive at another conclusion. In the list of drift-seeds given by Schimper [1]) 4 species (*Canarium, Chydenanthus* (*Barringtonia*) *excelsa, Pangium edule, Cynometra cauliflora*) are mentioned, which are no littoral plants. Of the 22 species of seeds and fruits (Cf. chapter VI) which Penzig in 1897 found washed ashore at Zwarte Hoek 9 did not originate from beach-plants. These seeds and fruits were too large for being driven by the wind into the interior. Such large seeds and fruits most easily strike the eye. But whoever has acquainted himself personally with the drift-zone will admit that the rubbish washed ashore may contain many small light seeds [2]) which easily escape the eye of the untrained investigator and may be driven by the wind into the interior to give rise there to a new vegetation. The many factors promoting or hindering the origin of such a vegetation cannot be discussed here. That abundance of ferns need not be the result of the large distance at which an island is situated from a plant-covered locality, is sufficiently shown by the fact that in Java the mountain-rainforests between 1000 and 2500 m. possess an extremely rich fern-vegetation. Constant moisture is of infinitely more importance than distance.

[1]) Schimper, Indo-Malayische Strandflora (1891), pp. 161, 162.
[2]) Cf. p. 46, footnote 1.

The two instances selected by Treub in corroboration of his argument were badly chosen. The group Juan Fernandez [1]) consists, besides of some small islands, of two larger ones, Mas-A-Tierra and Mas-A-Fuera, respectively 75 and 59 km^2. in size. Mas-A-Tierra, the highest top of which reaches about 1000 m., has a mean annual temperature [2]) of 15° C and a considerable rainfall, averaging 1004 mm. a year, distributed as follows:

Jan.	Febr.	March	April	May	June	July	Aug.	Sept.	Oct.	Nov.	Dec.
31	40	60	98	144	147	160	124	72	58	43	27.

Rainfall and temperature are sufficient for a rich forest-growth. The island, famous for its splendidly coloured flowers, is partly covered with rain-forest, though the most valuable indigenous trees have been exterminated. Epiphytical ferns (*Polypodiaceae* and *Hymenophyllaceae*) are plentiful in the rain-forest. Up till now 42 species of ferns [3]) have been found in Mas-A-Tierra, 51 in the entire group. There is only one Phanerogamic epiphyte, but numerous trees and shrubs, leading as a rule a terrestrial life, occur as accidental epiphytes in very moist localities. The undergrowth in the forest consists for by far the greater part of ferns and cell-cryptogams. About one third of the flora is endemic.

It is by no means proven that this abundance of ferns and scarcity of Phanerogams in the undergrowth of the forest is to be ascribed *solely* to seeds of Phanerogams not being introduced by cause of the *isolation* of the island. In every not too young rain-forest in Java one may observe that introduced plants either are entirely deficient or occur only in glades, clearings and along broad roads receiving a somewhat large quantity of light, but that they are unable to penetrate into the forest, though at a distance of a few meters numerous fructiferous specimens of these introduced plants may be found. The forest near Tjibodas on Mount Gĕdè strikingly illustrates this fact. Along the path to the cataracts many introduced plants occur along the forest-borders; in the forest proper adjacent to it not a single one.

The conditions of growth presented by Ascension (± 90 km^2.) differ greatly from those on Juan Fernandez. From the description given by Treub of the soil of Ascension may be concluded that

[1]) Cf. Schimper, Pflanzengeographie 2. Aufl. (1908) pp. 142, 520; Encyclopaedia Britannica, 11th Ed. XV (1911), p. 529; Winkler Prins, Geillustreerde Encyclopaedie X (1918), p. 221.

[2]) Data on the climate received by courtesy of Dr. Visser, subdirector of the Royal Magn. Meteor. Observ. Batavia.

[3]) Carl Christensen and Carl Skottsberg, The Pteridophyta of the Juan Fernandez islands, p. 45.

it is only fit for a kremnophytic vegetation. Whether this vegetation will consist of ferns or of other plants depends very much on the degree of moisture: a fern-association thrives only in not too dry localities. The mean annual temperature [1]) of Ascension [29 – 30° C at the coast, 24° C on the top (860 m.)] is very much higher than in Juan Fernandez, the amount of rain, on the contrary, is much smaller: it averages only 500 mm. a year. But there are years in which the rainfall on the island – locally at least – is much less than 500 mm. From 1863 till 1865 it averaged [2]) only 84 mm., distributed as follows:

Jan.	Febr.	March	April	May	June	July	Aug.	Sept.	Oct.	Nov.	Dec.
1	1	3	29	5	6	14	7	7	6	2	3.

Perhaps the latter record bears on another locality than the first. How this may be, it is certain that considering the nature of the soil, the very high temperature, the comparatively small rainfall and the impossibility of irrigation, the lower parts can bear only a meagre vegetation.

But another situation presents itself in the higher parts of the island. The top, popularly called Green Mountain, is very often wrapped in clouds which bring moisture and hinder evaporation. Here rises the only brook the island possesses, here such conditions are found as favor a kremnophytic vegetation: humidity and an unfertile soil; here the soil is clothed with a carpet of ferns. And here too does not exist a shadow of a proof that the scarcity of higher plants is due *only* to their seeds not being introduced.

Treub might have cited other instances of isolated islands, as St. Helena, the Canaries, the Hawaiian Islands. But all of these, bearing a rich vegetation of Phanerogams, disprove his hypothesis which therefore has to be rejected.

9. *When a new flora develops on a volcanic island which is in the same circumstances as Krakatao, the Phanerogams always will be preceded by Cryptogams because the latter, physiologically spoken, are less differentiated. In such-like localities the vascular cryptogams and especially the ferns, perform at present still the part they performed very often in earlier times, when they constituted the main vegetation of the earth.*

This conclusion too is rather vague. For there is no other island where as to climate, soil, environment and further circumstances,

[1]) Cf. Encyclopaedia Britannica, 11th Ed. II (1910) p. 716. Winkler Prins, Geillustreerde Encyclopaedie II (1915), p. 193.

[2]) These data on the rainfall received by courtesy of Dr. Visser, subdirector of the Royal Magn. Meteor. Observ. Batavia.

the same conditions prevail as in Krakatao. And so little is known of the vegetation of Krakatao in the first years after the eruption that a reliable comparison with other islands cannot be made, supposing the vegetation of these latter to be sufficiently known.

The predominance of ferns in the locality investigated by Treub is ascribed by him to the supposed fact that these plants are physiologically less differentiated than Phanerogams. By this expression he apparently meant to say that all ferns lay about the same claims on their substratum and environment. Taken in this wide sense Treub's thesis is undoubtedly wrong. In the tropics one finds among the ferns as well as among the higher plants terrestrial and epiphytical, halophobous and halophilous, heliophilous and heliophobous species. There are species which require a very moist substratum and species which prefer an only moderately moist one, species thriving only in soil rich in humus and species growing in humus-deficient soils. It is true that very many ferns do not require a fertile soil, that many are even typical kremnophytes. Hence such ferns could be numerous in the sterile locality investigated by Treub where Penzig 11 years afterwards found the vegetation practically unchanged. By the same cause the moist, higher parts of Ascension can bear a carpet of ferns. That certain ferns can predominate in unfertile, not too dry islands is caused partly by the modest claims they lay on the fertility of the soil, partly by their easy and rapid reproduction by stolons or spores, which are often formed in great quantities. Both of these factors are of immense value in the struggle for life. Among the Phanerogams occurring in the less dry regions of the Dutch Indies comparatively fewer kremnophytes are found and these often have less simple means of reproduction. When a new vegetation develops in a sterile constantly humid locality, the Phanerogams, therefore, at first may form only a small minority. Their number will increase as soon as the locality begins to offer favorable conditions for their growth and reproduction. Where these favorable conditions fail to come, the Phanerogams are unable to supersede the vegetation of Cryptogams which are better equipped for the struggle for life, and the succession predicted by Treub will never take place.

When we now proceed to sum up which *reliable* answers Treub's investigations have given to the questions put in the beginning of the Introduction, we arrive at the following conclusions:

1. It is unproven, not even rendered probable, that the original vegetation of Krakatao was totally destroyed by the eruption. In sheltered parts of localities which had *not* been covered with a *thick* layer of *hot* eruptive products and from where the covering layer soon disappeared (higher ravines, clefts in the basaltic rocks of Zwarte Hoek) a number of plants may quite well have survived the eruption.

2. It is unknown how soon after the eruption the pioneers of a new vegetation appeared on Krakatao, but it is a well-established fact that they were not long in coming. A year after the eruption a non-botanist meant already to see some plants; within 3 years after the eruption Treub, *from a great distance, saw rather big plants growing near the summit of the mountain.* Such plants had already been observed by others before him. After the rains had brought sufficient humidity and had washed away here and there the covering layer, probably in many localities at about the same time a new vegetation developed, consisting partly of remnants of the old one, for the rest originating from spores, seeds ands fruits saved or introduced in some way or other.

3. It is unknown which plants appeared first, which afterwards. Probably the algae, very moderate in their claims, soon made their appearance and spread rapidly; ferns and higher plants may have appeared in some places at the same time with the algae, in other places later or sooner. It is neither proven nor probable that everywhere the higher plants — as such — were preceded by ferns and these by algae. Of the succession of species or systematical groups nothing is known with certainty but that in a sample of the layer of algae, taken to Buitenzorg, *one* germinating fern-spore was found. To deduce from this fact the succession algae-ferns-Phanerogams seems rather risky.

4. As it has not been proven that the original vegetation of Krakatao was totally destroyed by the eruption and, moreover, next to nothing is known of this vegetation, it cannot be established which plants after the eruption have been introduced, which not. The tropical sea-shore plants

found by Treub on the beach had probably been introduced by ocean-currents, as the original littoral flora in all likelihood was quite destroyed. Whether the plants seen in 1886 in the interior had been introduced and, if so, in which manner, *is fully unknown;* it is possible but by no means proven. Which part man, animals, water and wind have taken in a possible introduction of plants into the interior *is quite obscure.* The investigations of Treub have not at all proven that any fruits, seeds or fern-spores were introduced by wind across the sea.

5. Treub found in 1886 on Krakatao already 3 plant-associations (see p. 70). The total number of species found by him in Krakatao amounted to 34. *But many more species may have been present in the island of which he investigated only a very small part.* Nothing is known on the flora of the rest. Unhappily this rest comprises the parts most important for the solution of the problem: those higher ravines on the south- and the eastern side, from which the covering layers had soon been removed. Consequently our knowledge of the vegetation in 1886 is very imperfect and can never serve as a base of comparison with the results of subsequent just as defective investigations, which were, moreover, often carried out in other localities.

6. We know but very little of the vegetation of the localities which Treub *did* investigate, even when we take for granted that this savant really saw and recorded *all* species of plants growing there, even very young seedlings, which is next to impossible. Treub does not state which ferns formed the main-vegetation and which were rare, neither does he record on which kind of soil the different species occurred. He says nothing of the degree of weathering, the amount of water and the temperature of the soil at different depths, nothing of the composition of the ground-water, nothing of soil micro-organisms, nothing of the extension of the subterranean and superterranean parts of the plants, nothing of the fauna and its relations to the flora. Therefore it is impossible to form a clear idea of the vegetation, one only obtains a misty general impression of the whole.

7. Summarized, so very little is known of the flora in 1886 that not a single reliable conclusion can be drawn, neither about the state at that time nor about the mode of development between 1883 and 1886. *All guesses made on this point, without a single exception, are unproved and unprovable, often even improbable.* The harsh words must be spoken: taken as a whole the botanical investigation of Krakatao in 1886 *was a failure;* because of its superficiality it could not be otherwise.

V. THE TRIP OF BOERLAGE TO LANG EILAND IN SEPTEMBER 1896.

(Jaarverslag 's Lands Plantentuin te Buitenzorg, 1896, p. 23. – Tijdschrift Koninkl. Nederl. Aardrijksk. Genootschap XIV (1897), pp. 118 – 122).

In the Annual Report of the Botanical Gardens of Buitenzorg over 1896, p. 23, Treub made the following statement (translated): „From a trip to Lang Eiland, situated near Krakatao, Dr. Boerlage, „then subdirector of the Botanical Gardens, brought a small collect- „ion of plants to Buitenzorg. This collection, together with another „one made by lieutenant Nolthenius, then stationed at the island, „gives a complete survey – as far as Phanerogams and Vascular „Cryptogams are concerned – of the young vegetation developed „on that island, after the eruption".

The same trip is mentioned in the Tijdschrift of het Koninklijk Nederlandsch Aardrijkskundig Genootschap (Journal of the Royal Dutch Geographical Society) XIV (1897), p. 118 seq. From an essay by J. J. A. Mulder the following may – in translation – be cited here:

From June 26th till July 2nd 1896 some officials of the Triangulation-brigade stayed on Krakatao charged with seeking a way to the summit which is about 800 m. high. Along different paths they tried to reach the top but all efforts failed. Notwithstanding very much trouble they were not able to get higher than a few hundred meters. The entire mountain [1]) is covered with a layer of ashes some tens of meters thick, into which deep ravines have been eroded by the rains. When ascending the mountain the visitors followed the narrow ridges between the ravines but all of these appeared to be broken at one or more spots by slides which could not possibly be crossed by means of the ladders the visitors carried with them. Even if these had been equipped much better the dangers would have been enormous, because everywhere the loose materials threatened to come down. As they did not intend to make a record of break-neck tours they stopped their efforts.

Instead of on Krakatao a pillar was erected on Lang Eiland, at a height of about 130 m. on the ridge extending throughout the island. There the difficulties were by far not so great.

Towards the close of August Lang Eiland was occupied as a

[1]) One should bear in mind that only the lower parts of the island were reached, so that this statement only holds good for these lower parts.

station for goniometric observations. The rains did not come, the Westmonsoon set in extraordinarily late . . . . In December heavy showers fell . . . . about the midst of January the observations were completed [1]).

Life on the island was very disagreeable in every respect. The days were very hot: in the dwelling-hut the temperature for days at a stretch rose as high as 34° C; by night it did not descend beneath 30° C; the sand heated by the sun had at noon a temperature of somewhat more than 60° C. Drinking-water had to be supplied from Batavia. For this purpose a steamer visited the island twice a month. There was no living water; *a well which had been sunk yielded water that, on chemical analysis, proved to be unfit for drinking-water because of its high percentage of organic substances.*

The vegetation on the little weathered layer of ashes was still in a stage of development. The trees were young, they were mainly Casuarinas which here and there began to form small groves; the most common plant was the grass *Saccharum spontaneum L.* Consequently there was but little shadow. In October Dr. B u r c k, the former sub-director of the Botanical Gardens at Buitenzorg and his successor Dr. B o e r l a g e paid a visit to the island in order to examine its flora.

The animal kingdom was represented by some varans (*Varanus salvator Laur.*), which apparently felt quite at ease here in the sea-water, and moreover by some birds and insects.

On the beach pumice was found in plenty. For the rest everything was covered with ashes; the sides of the hill-ridge were cleft by innumerable ravines, formed by erosion, with perpendicular walls of 40 and 50 meter high. These ravines were very peculiar; their sides were covered with a thin layer of algae, protecting them from being washed off by rain-water. Slides were formed from beneath by the action of water which had penetrated into the layer of ashes. – The bottom of the ravines was in the broader parts, at the entrance f. i., usually flat and almost horizontal.

On entering a ravine the visitor, secluded from the outer world by the high meandering ash-walls, soon got pervaded by an oppressive sensation of forlornness. Here and there on the bottom entirely decayed fragments and stumps of trees were lying, the sad remains of the heavy forest which formerly covered the entire island. Moreover the visitors found rapilli and, sometimes high up in the

---

[1]) But in March 1897 officials of the Triangulation brigade were still – or anew – living on Lang Eiland. Cf. P e n z i g in Ann. Jard. Bot. Buitenzorg XVIII (1902) p. 105.

ash-walls, rather big stones [1]), doubtless ejected by the crater of Krakatao during the eruption.

So far M u l d e r. — B o e r l a g e did not make his trip in October but a few days earlier, Sept. 29th 1896. As appears from the labels accompanying some of his specimens [i. a. *Nephrolepis exaltata Schott* and *Pteris vittata L.*] he reached the top of the island to which the coolies of the Triangulation Brigade had cleared a path. The collection, which is preserved in the Buitenzorg Herbarium, is rather large: it consists of at least 132 specimens but frequently several of these belong to the same species; the total number of species therefore is probably [2]) rather less than 100. I cannot share T r e u b's opinion that this collection gives a complete survey of the Phanerogamic and fern vegetation of the island. On the contrary, data on habitat, altitude and frequence as a rule being lacking (only a few times the labels bear, besides the collecting-number, the word *ravine* or *top,* written by a native assistant) it is of extremely little oecological or plant-geographical value. With the aid of the collection one can make a list of names of the plants found, but it gives by no means an idea of the mode of composition of the flora at that time. Moreover it cannot be but very incomplete. *It is absolutely impossible* to accurately investigate in a single day the innumerable ravines and ridges of Lang Eiland. For this purpose weeks, nay months are required. B o e r l a g e's investigations were probably carried out on the sea-shore near the landing-place and along the path leading from there through ravines and over ridges to the pillar. Taking into account that the perpendicular ravine-walls on Lang Eiland consisted of loose ashes, that they reached up to 50 m. and therefore could not be investigated without special equipment and at the cost of very much time, moreover that the young vegetation of the island must have consisted, besides of a number of common species, of several scattered accidental ones, it is obvious that B o e r l a g e's collection could not nearly comprise all species occurring in Lang Eiland. And even if it had done so, it would have been of very little value in the absence of all oecological data.

It is noteworthy but easily accounted for that B o e r l a g e collected on Lang Eiland some species [3]) which were not found

---

[1]) Cf. also p. 23, footnote 2.

[2]) No list of these species, to my knowledge, being extant, I have not succeeded in finding them all back. For this paper they are of very little importance.

[3]) i.a. *Blumea chinensis D. C.,-Blumea lacera D. C.,-Erigeron linifolius Willd.* [4]),- *Pteris vittata L.* [4]).

[4]) This species was found by P e n z i g only on Krakatao.

back half a year afterwards by Penzig and his companions who collected only on the beach. Of course these species were not recorded by Penzig in his list of plants found on Lang Eiland.

The collection made by Nolthenius is not very extensive and does not contain any oecological data. Hence it is practically valueless for the study of the development of the new flora.

The collections made in 1896 by Boerlage and Nolthenius on Lang Eiland, though preserved for upwards of 30 years in the Buitenzorg Herbarium and mentioned by Treub in the above-cited Annual Report, were never recorded in the lists of plants of the Krakatao-group published by subsequent authors. For the reasons set forth on p. 35 I refrain from further discussion of these collections and confine myself to stating that among at least 132 specimens collected by Boerlage there are to my present knowledge only 4 species of ferns (See p. 33, footnote 1). All of these are kremnophytes, very common in Java.

VI. THE EXPEDITION OF PENZIG, TREUB, RACIBORSKI, BOERLAGE AND CLAUTRIAU TO KRAKATAO, LANG EILAND AND VERLATEN EILAND IN MARCH 1897.

[Ann. Jard. Bot. Buitenzorg XVIII (1902), p. 92 seq.]

On March 17th 1897, half a year after Boerlage's and Burck's trip to Lang Eiland the 3 islands of the Krakatao-group [1]) were visited by Penzig, Treub, Raciborski, Boerlage and Clautriau. They were assisted by Pa Idan, a most excellent native collector, now already dead for many years, who by his phenomenal knowledge of the flora of the surroundings of Buitenzorg has rendered great services to the many foreign botanists who visited the Buitenzorg Botanical Institute.

Five years after his visit to Krakatao Penzig gave a survey of the results obtained, using thereby notes and photos avowedly originating from Boerlage and Clautriau who had already died by that time. From this survey it appears that the investigation carried out by Penzig and his companions suffered from the same defects as that of Treub. They too took a priori for granted that the former vegetation of Krakatao must have been entirely destroyed [2]) by the eruption of 1883; they too investigated only a very small — and not the most important — part of the island; the time they could give to their exploration — half a day only — was altogether insufficient for a close investigation of the vast and much accidented locality; they [3]) too lacked a sufficient knowledge of the Indian flora and its behaviour towards the factors of habitat and climate; they too

---

[1]) This excursion to Krakatao was preceded by a trip to the very small island Pulu Merak, lying off the north-east point of Java. This fact is mentioned here only because Penzig wrongly took his companions and himself for the first botanists who ever visited this island. But in 1862 already Miquel (Flora Ind. Batav. Sumatra, passim) recorded several plants found by Teysmann on Pulu Merak.

[2]) Cf. Ann. Jard. Bot. XVIII (1902), p. 103: Verlaten Eiland, „nur wenige Kilometer „vom Krakatau entfernt, hat zur Zeit der denkwürdigen Eruption *das Schiksal des letzteren „getheilt* [4]): die Gluth verdorrte und verbrannte jegliche Vegetation; Lava-, Bimsstein- „und Aschenmassen begruben unter meterhohen Schichten jede Spur thierischen und „pflanzlichen Lebens".

[3]) Penzig stayed only 4 months (Nov. 23rd 1896 till April 1st 1897) in Java and devoted this time mainly to mycological studies. But even if he had exclusively and assiduously studied Ferns and Phanerogams he could not possibly have obtained the knowledge necessary for a thorough phyto-geographic investigation for which a preparatory study of many years is required.

[4]) Italics by me.

paid little or no attention to oecological and biological relations [1]; in short, this trip may not be called a scientific oecological investigation made by experts, but was a mere excursion of persons interested into the problem but not seriously trying so solve it. Necessarily the scientific results were few. The very short descriptions of the vegetation found in Penzig's paper give but scanty and incomplete data about habitat and frequency of a small part of the species collected; several of these are passed over in silence; often it is not even mentioned whether a plant was found on the beach or in the interior. Penzig, however, was fully aware of the shortcomings of his investigations; he admits that the fact of a species not having been *found* by him or his companions is by no means to be considered proof that it did not *occur* then on the island.

The locality explored [2]) was about the same as that visited by Treub 11 years before (see p. 36). The rest of the island, including the so important ravines on the east- and south-east side, was not visited. As might be expected the investigators first explored the vegetation of their landing-place, the level sandy beach near Zwarte Hoek at the base of the rupture. This beach was still narrow though it had possibly increased in size after Treub's visit. Penzig and his companions found there a well-developed *Pes-caprae*-formation, mainly consisting of *Ipomoea pes caprae Sw.* and *Canavala rosea D. C.* [3]) both common littoral plants in the Dutch Indies; their far-creeping stems often form a loose net-work on the sandy soil [4]). Between these occurred *Vigna marina Merr.* [5]), a twining, or — if there are no supports to twine around — creeping Papilionacea, in Java very common on sandy beaches. Two littoral grasses with long creeping stolons were also present, *Spinifex littoreus Merr.* [6]) and *Ischaemum muticum L.*; both grasses often form a dense vegetation on the beaches of the Dutch Indies; the first only in very sunny, sandy localities deficient in humus; the second also in shadowed and less

1) Exception should be made for Raciborski († 1917) whose most excellent work, Die Pteridophyten der Flora von Buitenzorg (1898), everywhere gives proof of accurate observation and clear insight.

2) Penzig [Ann. Jard. Bot. Buitenzorg XVII (1902), p. 97] gives the following description of the landing-place: „An der westlichen Hälfte der Nordseite der Insel „wo wir landeten, ist eine schmale Strecke flachen Strandes gebildet". — This proves (see also our map) that he can have visited only the locality near Zwarte Hoek as mentioned by me. Treub also landed on that spot in 1886. From this beach the interior of Krakatao can be reached only over the rocks of Zwarte Hoek.

3) Named by Penzig *Canavalia obtusifolia D. C.*

4) A good reproduction of a photo is found in Ernst, Neue Flora Vulkaninsel Krakatau (1907), plate IV, fig. 7.

5) In Penzig's paper recorded under the names of *Vigna lutea* and *V. luteola*.

6) Named by Penzig *Spinifex squarrosus L.*

poor habitats. Furthermore 2 Cyperaceae, *Cyperus pennatus Lmk.* [1]) and *Fimbristylis spathacea Roth,* tussocky, salt-resistent herbs which prefer a firmer substratum than the plants mentioned before; they are mainly found on a clayey, rather moist soil (either saline and often so, or not) behind the beach, frequently rather far (a kilometer and more) remote from it, but they may also occur on the older, firmer parts of a broad beach. Furthermore (probably) *Euphorbia atoto Forst* [2]), a deeply rooting halophile herb, common on many loose, sandy beaches and very conspicuous by its glaucous hue. Finally *Scaevola frutescens Krause* [3]) and *Pandanus tectorius Sol.* [4]), robust shrubs or (*Pandanus*) small trees of sandy beaches, both very common in Java. *All these species were represented by very numerous specimens,* which totally covered part of the beach and flowered and fruited freely, so that Penzig thought they would certainly hold out in the locality where they had established themselves.

Besides these plants mentioned by Penzig as growing on the beach, some other littoral plants, belonging to the *Barringtonia*-formation, must have occurred in this locality as appears from the list of plant-names given by Penzig. Unhappily in this list data on habitat and frequency of these species are entirely lacking, but it is much more probable that they grew on the beach, which is their natural habitat, than in the interior, which is not.

Penzig's list of plants found on Krakatao contains the following littoral or halophilous species:

Pandanaceae.

1. Pandanus tectorius Sol. (P. littoralis Jungh.). See p. 43, No. 4.

Gramineae.

2. *Ischaemum muticum L.* See p. 89.

3. *Spinifex littoreus Merr.* (Sp. squarrosus L.). See p. 89.

The reproduction of these two grasses is effected, besides by fruits, also by far creeping superterranean stolons [5]).

---

[1]) Probably this is the species named by Penzig *Cyperus digitatus Roxb.* The latter is no littoral species at all.

[2]) Penzig mentions the plant, which he calls a common inhabitant of sandy sea-shores, under the name of *Euphorbia pilulifera L.* (= *Eu. hirta L.*), but this name is most probably wrong. *Euphorbia hirta L.* is not a common littoral plant but a weed of fields, lawns, roadsides and gravel-covered compounds. It was never found back on the spot in question at subsequent excursions; *Euphorbia Atoto* on the contrary was found there in 1905, 1906 and 1908.

[3]) Named by Penzig *Scaevola Koenigii Vahl.*

[4]) Named by Penzig *Pandanus littoralis Jungh.*

[5]) On a stolon of 5 m. length of *Ischaemum muticum* I found in 1911 on the beach of Tji Langkahan in Western-Java 122 young individuals.

Cyperaceae.

4. *Cyperus pennatus Lmk.* (In Penzig's paper wrongly named C. digitatus Roxb.).

5. *Fimbristylis spathacea Roth.* (See p. 90).

These two species are reproduced by seeds only.

Casuarinaceae.

6. *Casuarina equisetifolia Forst.*

Fast growing tree of level or gently sloping, well-pervious beaches rich in lime and lacking in humus and similar localities close back of these. Kamerling [1]) found tubercles on its roots, which he *thinks* may be inhabited by nitrogen-fixing bacteria in the same way as those of many *Leguminosae*. In Java this tree has been found only in a few localities (i.a. on the beach of Tji Lintjing, east of Tandjong Priok, which is very rich of shells), but on several coral-islands in the Java-sea it is rather numerous. The fruits, which burst when ripe and then easily let escape the winged seeds, can remain floating on sea-water for a few days; the seeds themselves possess also floating-power [2]). Over land the seeds may be spread by wind, across the sea by sea-currents. When the tree has attained a certain age it emits from its roots many young shoots which often form a small grove around an old trunk. If, by some cause or other, the substratum grows too rich of humus, *Casuarina*, when not protected by man, gets gradually superseded by other trees [3]).

Lauraceae.

7. *Cassytha filiformis L.*

Twining or partly decumbent, much branched, green or brownish-orange, polyphagous, long-lived parasite [4]), often

---

1) Z. Kamerling, Over het voorkomen van wortelknolletjes bij *Casuarina equisetifolia* in Natuurkundig Tijdschrift Ned.-Indië LXXI (1911), p. 20.

2) On the Keeling-islands Guppy found the seeds in the drift-zone.

3) Such is, locally at least, since some years happening on the islands of the Krakatao-group. Docters van Leeuwen [5]) states that in 1921 on Verlaten Eiland the old forest of *Casuarina equisetifolia L.* in which the trees grew wide apart and gradually perished, was in course of being replaced by a mixed forest and that the *Casuarina*-forest on Krakatao „was in several places mostly already fairly old „with few rejuvenated parts".

4) A specimen in the Buitenzorg Botanical Gardens had in 1924 reached an age of upwards of 20 years.

5) W. Docters van Leeuwen, The Vegetation of the Island of Sebesy, situated in the Sunda-Strait, near the Islands of the Krakatau-group; in the year 1921, in Ann. Jard. Bot. Buitenzorg XXXII (1923), pp. 158, 159, 160.

forming a dense net-work over grasses, shrubs and small trees. In Java it is very common in the *Pes-caprae*-formation and in young *Barringtonia*-formations; rarely it is found in thickets in the interior. The ripe fruit is a whitish spurious berry consisting of the rather thin fleshy perianth including the dry, one-seeded fruit proper. Both Schimper and Ernst [1]) assert that these spurious berries are eaten by birds; Guppy [2]) found them in the gizzard of frugivorous birds. The fruits which have not been eaten by birds finally fall off, still enveloped by the then withered perianth; they possess a well-developed floating power because of the fruit-cavity not being entirely filled up by the seed. At Buitenzorg they remained floating for somewhat more than a week on a $3^1/_2$ % solution of NaCl. Consequently the fruits of this species may be transported by water, but probably the dispersal is, sometimes at least, brought about by birds also. If such is really the case, the rare occurrence of the plant in the interior far above high-tide mark [3]) finds a natural explanation. But only semination of seeds found in bird-droppings can furnish the required proof.

Leguminosae.

8. *Canavalia rosea D.C.* (In Penzig's paper named *C. obtusifolia D. C.*). See p. 89.

A rather common plant of the *Pes-caprae*-formation, with far-creeping, never twining stems. It has often been confused with *C. obtusifolia D. C.*, which, according to Prain, is a twining species; it has also been confounded with *C. lineata D. C.* The reproduction is effected as well by ramification of the creeping stems as by seeds. The pods burst open when ripe; the seeds [4]) possess floating power.

---

1) Schimper, Indo-Malayische Strandflora (1891), p. 188, — Ernst, Neue Flora Vulkaninsel Krakatau (1907), p. 11. It seems that in the Buitenzorg Botanical Gardens the fruits, though plentifully produced and by their whitish hue conspicuous from afar, are not much sought after by birds. Of course this by no mean proves that such is also the case on the beach.

2) Hemsley, Challenger Reports, Botany I, pp. 43 – 44.

3) In 1925 Dr. J. Coert and I found a specimen of this plant (together with several specimens of *Acrostichum aureum L.*) in thickets surrounding a salt spring near Ngantang (west of Malang, Java) at an altitude of somewhat more than 600 m. above sea-level. Another specimen was found in 1911 at Bajah in West-Java, also far from the sea. Dr. C. van Dillewijn and I found yet another one in 1927 on the dry calcareous hills of Grisee some 50 meters above sea-level and about 4 km. from the sea.

4) Schimper, Indo-Malayische Strandflora (1891), p. 166.

9. *Vigna marina Merr.* (named by Penzig Vigna lutea D.C. and V. luteola D.C.). See p. 89.

The reproduction of this herb takes place exclusively by seeds. The black pods burst open when ripe and then or shortly afterwards let escape the seeds. The cotyledons of quite ripe seeds include between them [1]) an air-cavity by the presence of which the seeds possess floating power.

Euphorbiaceae.

10. *Euphorbia atoto Forst.* (See p. 90).

The reproduction of this plant is brought about by seeds which are flung away the moment the capsules burst. With ripe seeds the embryo does not entirely fill up the testa [2]); in consequence they possess floating power. In a $3^1/_2$ % solution of NaCl they remained floating for 4 à 5 days [3]).

Malvaceae.

11. *Hibiscus tiliaceus L.*

A tree which is frequent in the *Barringtonia*-formation; the spontaneous reproduction is effected by seeds. Because of the embryo not entirely filling up the testa [4]), the seeds possess floating power.

Lecythidacaceae.

12. *Barringtonia asiatica Kurz* (B. speciosa Forst.). See p. 43, No. 5.

Combretaceae.

13. *Terminalia catappa L.* See p. 39, No. 2.

Apocynaceae.

14. *Cerbera manghas L.* (Cerbera odollam Gaertn.). See p. 46, No. 3.

Convolvulaceae.

15. *Ipomoea pes caprae Sw.* See p. 48, No. 7 and p. 89.

Rubiaceae.

16. *Morinda citrifolia L.*

A small tree, frequent in the *Barringtonia*-formation. The reproduction is effected exclusively by fruits, which possess

---

1) Schimper, Indo-Malayische Strandflora (1891), pp. 163, 164.
2) Schimper, Indo-Malayische Strandflora (1891), p. 163.
3) Schimper, Indo-Malayische Strandflora (1891), p. 165.
4) Schimper, Indo-Malayische Strandflora (1891), pp. 163, 165.

floating power, but because of their great succulence soon rot. The seeds contain a large air-chamber (compared by Schimper [1]) to an air-bladder), hence possess floating power.

Goodeniaceae.

17. Scaevola frutescens Krause (Sc. Koenigii Vahl). See p. 49, No. 9.

Compositae.

18. Wedelia biflora D. C. (mentioned in Penzig's paper as Wedelia asperrima Boerl. and W. scabriuscula Boerl.). See p. 49, No. 1.

It is quite possible that besides these 18 species a number of non-halophile ones were growing on the beach examined by Penzig, especially on the older parts, which had become salt-free by leaching and therefore, oecologically spoken, did no more belong to the beach proper. *Saccharum spontaneum L.*, a halophobous species, has certainly done so as appears from the legend under a photo [2]) reproduced by Penzig on which this grass is distinctly recognizable. In 1906 and 1908 *Saccharum spontaneum* was in this locality one of the most common [3]) grasses on the oldest parts adjacent to the rupture. Hereafter I shall revert to this grass.

Consequently we may safely assume that Penzig on the beach S. E. of Zwarte Hoek found at least 19 species and that 11 of these (No. 1, 2, 3, 4, 5, 8, 9, 10, 15 and 17 of the list above, likewise *Saccharum spontaneum L.*) were quite common there.

Treub had investigated in 1886 the same part of the beach as Penzig and his companions did in 1897. This part being not very large [4]) and causing no difficulties to the investigation, one is perhaps warranted to believe that both investigations were sufficiently accurate to admit a comparison of the results, though it should never be forgotten that young plants may be easily overlooked or not recognized and that in the same locality 5 botanists, as a rule, will find more than one, especially if that one is no florist and the 5 others are assisted by a very able plant-collector [5]).

Of the 18 littoral or halophilous plants found in 1897 by Penzig

---

1) Indo-Malayische Strandflora (1891), p. 165 tab. VII, fig. 26 b.c.

2) Ann. Jard. Bot. Buitenzorg XVIII (1902), p. 100.

3) Cf. Ernst, Neue Flora Vulkaninsel Krakatao (1907) tab. IV, fig. 7, back-ground.

4) According to the map in the „Handelingen van het Congres" of which plate III is a reduced simplified copy, it is about 750 m. in length and at the broadest part (in 1908) 150 m. wide. In 1897 it was probably narrower, in 1886 perhaps very narrow. Neither Treub nor Penzig give data on this point.

5) I have not been able to ascertain whether Treub in 1886 was accompanied by Paldan.

and his companions, but 4 (No. 14, 15, 17 en 18) had been found with certainty by Treub in 1886, the last-cited species, however, in the interior only (See p. 49, No. 1). If the two Cyperaceae found by Treub on the beach were identical with those found by Penzig (No. 4 and 5), the total in 1886 amounts to six. Finally Treub found in that year 3 littoral plants (*Calophyllum inophyllum L.* — *Hernandia peltata Meissn.* — *Tournefortia argentea L. fil.*, the last species, however, being mentioned by him for the interior only) which were not found back in 1897. Whether in consequence of unfavorable circumstances (See pp. 44, 45) these littoral plants had disappeared in the time elapsed between the two investigations or were overlooked in 1897, nobody can tell; the two first were found back in 1906 in the same locality, the third not.

Considering that Treub took part in the expedition of Penzig and surely has spoken to the latter on the differences between his impressions in 1886 and 1897, taking moreover in account that at both expeditions rare or inconspicuous plants may have been overlooked, yet one gets convinced that (as might be expected from a very young vegetation in a favorable locality) the flora of the level beach S.E. of Zwarte Hoek in the 11 years between the two excursions had grown richer, not only of species but especially of specimens. The species found by Treub in 1886 were represented by few and young specimens [1]) which did not yet form a dense vegetation. Penzig on the contrary expressly states [2]) of a number of species that in 1897 they occurred in numerous flowering and fruiting specimens and densely overspread the beach. The *Pes-caprae*-formation, which during the visit of Treub was just beginning to develop, had 11 years afterwards grown very much richer, but yet it was by no means fully developed, for many members of that formation (and of the *Barringtonia*-formation which gradually develops in it when there are no disturbing influences as f.i. decrease of the beach, too great mobility of the sand, destruction caused by man, cattle, fire) were still [3]) lacking or at least remained unobserved. Assuming that the

[1]) Ann. Jard. Bot. Buitenzorg Im Serie VII (1888), p. 217: „J'ai trouvé sur la plage „quelques jeunes pieds. . . ."

[2]) Ann. Jard. Bot. Buitenzorg XVIII (1902), p. 98: „Von den genannten Arten sind „schon sehr zahlreiche, den Strand z.T. völlig bedeckende Exemplare vorhanden, welche „reichlich blühen und fruchten".

[3]) In the same locality the following 12 „new" littoral plants were found in 1908: *Albizzia retusa Bth.* — *Calophyllum inophyllum L.* — *Clerodendron inerme Gaertn.* — *Derris heterophylla (Willd) Backer* (D. uliginosa Bth). — *Desmodium umbellatum D. C.* — *Dodonaea viscosa Jacq.* — *Erythrina variegata L.* var. *orientalis Merr.* — *Pithecolobium umbellatum Bth.* — *Premna integrifolia L.* — *Sophora tomentosa L.* — *Vitex paniculata Lmk.* — *Ximenia americana L.*

investigation in 1886 was carried on with sufficient accuracy, we arrive at the conclusion that in the 11 years elapsed between the two excursions 12 new littoral plants (No. 1, 2, 3, 4, 5, 6, 7, 8, 9, 10, 11, 12, 13 and 16 of the above list), moreover *Saccharum*, had established themselves *in the locality investigated* and several of these species (1, 2, 3, 8, 9, 10 and Saccharum) occurred in numerous specimens. All these new littoral plants (*Saccharum* is *no* littoral plant) have seeds or fruits fit to be carried along by sea-currents; of 3 of these species (No. 1, 12 and 13) Treub had in 1886 already found the seeds or fruits washed ashore.

This, of course, by no means proves, nor even renders it probable, that these new species had *reached Krakatao after 1886.* Treub examined in 1886 only a small part of the beach, at best one sixth of the whole [1]. *As to the flora of the rest of the beach in that year we are in complete ignorance.* All the littoral plants found in 1897 at Zwarte Hoek may quite well have grown in or before 1886 elsewhere on the beach of Krakatao [2]. Seeds or fruits of littoral species are often carried along for some distance by a current running parallel to the beach and then washed ashore. One specimen may – circumstances being favorable – in this way spread its posterity far over the strand. I need scarcely argue that if only part of a beach of an island is investigated and one might find there seeds, fruits or young seedlings but no large plants, this by no means furnishes proof that the species in question has but recently reached the island or has newly established itself. In order to decide on this point one has to examine thoroughly the *entire* beach. Such an investigation, which would take very much time, has never yet been carried out on Krakatao. Therefore it is impossible to ascertain at which time the littoral plants found there have first established themselves. This part of the problem will for ever remain unsolved.

Besides plants on the beach Penzig an his companions found in the drift-zone at Zwarte Hoek 22 seeds and fruits [3] mentioned by him under a specific name, viz. (recorded under the present names arranged in alphabetic order): *Barringtonia asiatica Kurz* (=B. speciosa Forst), – *Blumeodendon tokbrai Bl.* (= Elateriospermum tokbrai Bl.),—

[1]) See page 71, footnote 1.

[2]) The statement of Penzig [Ann. Jard. Bot. Buitenzorg XVIII (1902), p. 97] that „an fast allen anderen Seiten des Krakatau die felsige Küste steil und ganz unzugäng- „lich ins Meer abstürzt", does not hold good for the east- and south-eastern side of the island where a well-developed beach is present, the flora of which was first examined in 1905 and more intensively in 1906.

[3]) I have not been able to find these seed and fruits in the Buitenzorg collections. Hence it is impossible to control the determinations, however desirable this would be.

Caesalpinia crista L. (=C. bonducella Flem.), – *Cerbera manghas L.* (=C. odollam Gaertn.), – Cocos *nucifera L.*, – *Derris heterophylla Backer* (=D. uliginosa Bth.), – *Dillenia aurea Smith*, – *Entada phaseoloides Merr.* (=E. scandens Bth.), – *Gluta renghas L.* (= Gl. benghas L.), – *Guettarda speciosa L.*, – *Hernandia peltata Meissn.*, – *Inocarpus edulis Forst.*, – *Mangifera foetida Lour.*, – *Mangifera indica L.*, – *Nephelium juglandifolium Bl.* (= N. altissimum T. et B.), – *Nipa fruticans Wurmb*, – *Pandanus tectorius Sol.* (=P. littoralis Jungh.), – *Quercus daphnoidea Bl.*, – *Quercus pseudomolucca Bl.*, – *Scaevola frutescens Krause* (=Sc. koenigii Vahl.), – *Terminalia catappa L.* and *Xylocarpus moluccensis Roem.* (= Carapa moluccana Penzig = C. moluccensis Lamk.). – Part of these seeds may have come from the island itself, the rest must have been introduced, probably by the sea. As nothing is known on the flora of the higher ravines in 1897 (see p. 27), one would fruitlessly try to investigate which of the seeds or fruits might have originated from the interior of Krakatao [1]). Such fruits and seeds could hardly have come over from the interior directly to the drift-zone of Zwarte Hoek, because in 1897 behind that drift-zone lay already a strip of level old beach which was never reached by the waves and bore a rather dense vegetation. No ravines come out on this beach, which is secluded from the interior by a steep wall. The seeds and fruits supposed to come from the interior had to follow a circuitous way; they *may* have been carried to the sea by rain-water flowing during or shortly after heavy showers rapidly down the ravines on the east-, south- and west-side of the cone; the sea *may* have conveyed them round the island to the beach of Zwarte Hoek. But all these suppositions are mere guesses.

It is *certain* that among the seeds and fruits found in 1897 in the drift-zone at Zwarte Hoek there were 9 that could *not possibly* contribute to enrich the flora. P e n z i g [2]) – assuming that the determinations were correct – is wrong in considering the seeds and fruits recorded by him as originating without exception from littoral plants. *Blumeodendron tokbrai Bl.*, – *Dillenia aurea Smith*, – *Inocarpus edulis Forst.*, – *Mangifera foetida Lour.*, – *Mangifera indica L.*, – *Nephelium juglandifolium Bl.*, – Quercus *daphnoidea Bl.* and Q. *pseudomolucca Bl.*, are no littoral plants and had, even admitting that their seeds had retained their germinative power, which seems not to have been examined – no chance at all to live in the locality

[1]) None of the species named was afterwards (at least until 1919) found in the interior of Krakatao barring Cocos *nucifera* and this only rather close to the coast.
[2]) Ann. Jard. Bot. Buitenzorg XVIII (1902), p. 109.

reached. The fruits or seeds of these species are too heavy to be carried by wind farther inward; the beach in question is moreover almost entirely secluded from the interior by a steep wall (see p. 36). No chance to live had also the fruits of *Nipa fruticans Wurmb*, a palm of brackish swamps and river-mouths.

Specimens of 5 of the 13 remaining species (*Barringtonia asiatica Kurz*, – *Cerbera manghas L.*, – *Pandanus tectorius Sol.*, – *Scaevola frutescens Krause* and *Terminalia catappa L.*), were in 1897 already growing in the locality examined. *Hernandia peltata Meissn.* had been found there in 1886 by Treub. It was not found back in 1897, but Penzig himself admits that it may have been overlooked. *Caesalpinia crista L.* and *Guettarda speciosa L.* are quite common on many sandy beaches; the first had in 1919 not yet been found in Krakatao but was collected in 1906 and afterwards on Verlaten Eiland; possibly it occurs also somewhere on the beach of Krakatao; the second has at present been found in all three islands of the Krakatao-group. *Entada phaseoloides Merr.* (often mentioned under the names of *E. Pursaetha D. C.* or *E. scandens Bth.*) is a gigantic liana of the interior; its large seeds are apparently sometimes carried by rivers to the sea, at least they are not rarely found washed ashore and sometimes they germinate on the beach. In 1897 a specimen of this species was collected on the beach of Verlaten Eiland, in 1908 on that of Lang Eiland; it may quite well grow here or there on the beach of Krakatao. *Derris heterophylla* (*Willd.*) *Backer*, mentioned by Penzig under the synonym *D. uliginosa Bth.*, is a creeping or climbing shrub, which is very common in many parts of the mangrove; at present it has already been found on all the islands of the Krakatao-group. For *Gluta Renghas L.* see p. 45; young specimens of this tree of a man's height were collected in 1908 on Verlaten Eiland and Lang Eiland in low and rather old parts of the beach. *Xylocarpus moluccensis Roem.* is an inhabitant of the mangrove [1]); it may have grown in 1897 somewhere on the beach of Krakatao. Another species of the same genus, likewise a component of the mangrove, *Xylocarpus Granatum Koen.*, was found in 1906 on the south-eastern beach of Krakatao (See

[1]) Docters van Leeuwen [Handelingen Eerste Ned.-Indisch Natuurw. Congres (1920), p. 42.-Ann Jard. Bot. Buitenzorg XXXI (1921), p. 112] wrongly names *Lumnitzera racemosa Willd.* and *Excoecaria agallocha L.* as the *only* mangrove-plants of the Krakatao-group. Other components of the mangrove, given in the list published by himself and found there already before 1919 are: *Acrostichum aureum L.*,- *Derris heterophylla* (*Willd*) *Backer* and *Xylocarpus Granatum Koen.* Though the beaches of Krakatao are unfit for the development of a mangrove-*vegetation*, specimens of mangrove-components can grow in favorable spots. Of such mangrove-components 5 had been found up to 1920.

chapter VIII). On *Cocos nucifera L.* I gave already my views on pp. 39 seq. Penzig found nuts washed ashore but does not mention to have seen germinating nuts or young trees.

On the beach at Zwarte Hoek Penzig and his companions found several seedlings which furnished proof that a great part of the seeds washed ashore were able to germinate. It is much to be regretted that Penzig does not mention the names of these seedlings which he rightly considered to be „instructive Beweisstücke", and which were collected with „besonderem Eifer". For the problem of revegetation they certainly were of much more importance than the seeds which had not yet germinated.

After having examined the beach the visitors penetrated into the interior of the island. In order to do so they had to ascend the rocks of Zwarte Hoek; everywhere else the beach is secluded from the interior by an unscalable wall. These same rocks were in 1886 examined by Treub. As they cover a rather large area, offer many difficulties to the investigation (see p. 29) and the whole exploration in 1897 was of but half a day's duration, they could be examined but very cursorily. Moreover, it is not at all certain that the part of the interior examined in 1897 was really identical with that of 1886, that the party followed quite the same route as Treub had done 11 years before. Referring to what I have said on pp. 5–7 on the great local differences a young vegetation may show I need scarcely argue the necessity of being very cautious when comparing the results of 1897 with those of 1886.

On the hills investigated and in the shallow valleys between, the vegetation, according to Penzig, consisted in 1897 almost exclusively of three tall grasses, viz. *Pennisetum macrostachyum Brongn.* (Gymnothrix elegans Büse, see p. 48, No. 8), *Saccharum spontaneum L.* and *Neyraudia madagascariensis Hook. fil.* var. *Zollingeri Hook. fil.* (in Penzig's paper mentioned under the name of Phragmites Roxburghii Steud.). In the dense jungle formed by these grasses other plants occurred only in scattered specimens. The 3 grasses named collectively by Penzig are by no means oecologically equivalent; in 1897 they were doubtless not present in almost equal numbers, neither nearly regularly dispersed. *Pennisetum macrostachyum Brongn.*, easily recognizable by its drooping, caudiform, white or yellowish spikes, is a perennial, tillering grass, $1^1/_2$–3 m. high, with a short, not very deeply buried rhizome; old specimens form small tussocks. This species grows only in sunny and often rather unfertile localities, frequently on steep slopes but also in crevices of rocks and such-like stony places. In Java, where it is often cultivated for ornamental purposes but is only rarely met with in a wild state, it never grows gregariously but always in scattered specimens. The specimens found in 1906 and 1908 on Krakatao, in localities which bore a dense vegetation as well as in those which bore only a meagre one, grew in the same manner and there is no reason to suppose that in 1897 the reverse would have occurred. On the mode of dispersion of this plant see p. 48.

In quite another way *Saccharum spontaneum L.* (Glagah, Kasò) behaves, a most variable species, generally 3–5 m. high, with large (30–100 cm. long) terminal flower-panicles, which during and after the anthesis (in Java especially from February till August; in other months flowering specimens are comparatively scarce) are very conspicuous [1]) by the long violet or white hairs clothing the bases of the spikelets. This rather deeply rooting, heliophilous grass is spread from West- to East-Java, from the plains up to an altitude of 1800 m. It is not a kremnophyte; though it does not require a fertile soil, it does not grow well in a very poor one. Wherever it occurs in great numbers the soil must be considered sufficiently fertile for divers agricultural plants. It needs rather much water to thrive well [2]); in regions where the east-monsoon is strong and the soil dries out to a great depth it is restricted to depressions of the ground and to water-sides, but where the east-monsoon, as a rule, is interrupted by repeated rains it also grows in other localities, especially in glades, on margins of forests, in thickets and young secondary forests, on ravine-sides and very often in old alang-fields. The occurrence of this grass in numberless specimens in the pumice- and alang-fields of Krakatao proves not only that in this island the somewhat deeper layers of the soil as a rule do not dry out very much, but also that the soil is not very poor. Because of the rhizome ramifying very much, creeping far and at short distances emitting new culms, old specimens of this grass finally form large and very dense bushes. In localities where the plant thrives particularly well these bushes may blend into vast jungles which can be penetrated into only with the aid of a chopping-knife.

---

[1]) See Ernst, Neue Flora Vulkaninsel Krakatau (1907), tab. VI, fig. 10.

[2]) Docters van Leeuwen [Handelingen Eerste Ned.-Ind. Natuurw. Congres (1920), p. 49; Ann. Jard. Bot. Buitenzorg XXXI (1921), p. 120] wrongly calls this plant a grass of *dry* soils. Just the reverse is true. One should not judge of the humidity of a soil by casting a look on its surface. See also what the excellent observer Junghuhn (Reisen durch Java, p. 216, 343) writes on the *Saccharum*-jungle on Mt. Galunggung and on the occurrence of *Saccharum* at the lake of Ngĕbĕl.

Besides by means of its rhizomes the grass easily and rapidly multiplies by its myriads of fruits. When these are ripe, the articulate branches of the panicle break into small fragments to which the fructiferous spikelets remain attached; finally only the naked main axis of the inflorescence with the bases of the lateral axes is left. In consequence of their being clothed with long hairs the fructiferous spikelets are easily dispersed by wind. Where the spikelets have fallen to the ground or on other plants they may be easily carried on high again by every strong gust of wind and may, in this manner, be conveyed over a distance of several hundreds of meters from the mother-plant. It is unknown how far the wind can carry them across the sea [1]) but it has not yet been proven that in a single flight they can traverse a distance of more than 20 km.

Junghuhn [2]) found in August 1837 a dense *Saccharum*-jungle on the gigantic mud-stream (lahar) which 15 years before (Oct. 1822) had flown from Mount Galunggung. This mud-stream is situated in a region where the 4 driest months of the year possess together 40 à 60 rain-days, hence, on an average, a rain-day alternates with but 1–2 dry days. Consequently there is no question of the lower soil-layers drying out very much, „der ganze Boden ist von Feuchtigkeit durchdrungen" [3]) and the opportunity for a *Saccharum*-vegetation to develop in that locality was extraordinarily favorable. The description given by Junghuhn [4]) of *Glagah* and the vegetation formed by it is so very vivid and at the same time so correct that I cannot resist the temptation to copy it here: „Mehrere andere eben so hohe oder „noch höhere Grasarten kommen hier und da, teils vereinzelt, dem „Alang alang untermengt, teils gruppenweise zerstreut darin vor . . . „vor allen andern die berüchtigte Glagah, *Saccharum spontaneum L.*, „dessen rohrartige fingerdicke Stengel mit ihren langen Blättern und „überhängenden Blütenrispen so hoch emporragen, dass Ross und

---

1) Spikelets dropped into the sea can no more be carried upwards by the wind. But it may be possible that after having been transported by sea-currents, they are washed ashore on Krakatao. After having dried on the beach they might be carried by wind into the interior. The gregarious growth of this grass on the older parts of the beach at Zwarte Hoek at the base of the rupture might be explained in this manner. But it is just as possible that *Saccharum* has reached this beach from the interior of the island. It is unknown how long fruits of *Saccharum spontaneum* can float on the sea without losing their germinative power. None of the authors on the new flora of Krakatao seems to have made any experiment in order to ascertain this. In accordance with the venerable recipe it was much more easy to have fruits introduced by a paper wind.

2) Junghuhn, Reisen durch Java (1845), p. 216 (under the name of *Saccharum Klaga*) and Java, 2nd Dutch Edition II, 158, 159 (under the name of *Imperata Glagah*).

3) Junghuhn, Reisen durch Java (1845), p. 216.

4) Junghuhn, Java, 2nd Dutch Edition I, 290,— German Edition I, 212.

„Reiter sich sicher darin verbergen können. Die Glagah (*Saccharum*) „wächst gewöhnlich gruppenweise im Alang (*Imperata*) und bildet dann „hier und da zerstreute Inseln, höhere Graswäldchen, die sich 8 bis „10, ja 12 Fuss hoch erheben, während die allgemeine Grasoberfläche „rings umher nur etwa 4 Fuss hoch ist. Zuweilen bildet sie aber auch, mit „Ausschliessung anderer Grasarten, Wildnisse für sich selbst, die „grosze Strecken einnehmen und schwieriger zu durchdringen sind, „als der dichteste Urwald, da man seine Pfade, Schritt vor Schritt, „in das dicht gedrängte Rohr mit den Hackmessern hineinhauen muss „und beim Straucheln Gefahr läuft, sich an den spitzen Rohrhalmen „wie auf Lanzen zu spiessen, die gewöhnlich einen Fuss hoch über „dem Boden schief abgehauen sind."

This description does not suffer from exaggeration. In 1906 Ernst and his companions found another dense *Saccharum*-jungle on the south-eastern side of Krakatao and Ernst vividly depicts [1]) how, after many fruitless exertions, the hope of crossing that jungle had to be abandoned: „Sengend heiss brennen die Strahlen der „senkrecht über uns stehenden Sonne auf uns nieder, abwechselnd „handhaben wir an der Spitze der kleinen Kolonne das Haumesser; „Schritt um Schritt kämpfen wir uns vorwärts in dem schattenlosen „Dickicht, aufwärts und wieder abwärts. Die Diener und Träger „mit dem notwendigsten Gepäck und den Pflanzenkisten vermögen „kaum zu folgen. Schiesslich müssen wir, wenn auch mit Wider- „streben, erkennen, dass die Erreichung der Hauptschlucht und des „Gipfels mit unserer Ausrüstung und in der kurzen, noch zur Verfü- „gung stehenden Zeit nicht möglich ist". In 1908 a gang of coolies of the Topographical Survey Service, when cutting a way through this same *Saccharum*-jungle, advanced in 4 hours not yet fully 300 meters [2]).

Such a dense growth of the grass is only possible where the roots almost everywhere can penetrate deeply into the soil, where the latter is sufficiently fertile and its lower layers are constantly humid. If either depth, fertility or moisture of the soil are insufficient, the *Saccharum*-vegetation is feeble and thin or entirely absent. A very reliable observation thereabout is mentioned by Fransen Herderschee [2]): „Suddenly we reach a broad ravine with a „stony bottom and low walls. It reaches down to near the coast and „has all the appearance of an old lava-stream. We can get on now

[1]) Ernst, Neue Vulkaninsel Krakatau (1907), p. 33.

[2]) Jaarverslag Topogr. Dienst Ned. Indië (Year-book of the Topographical Survey Service in the Dutch Indies) 1908, p. 155.

„without trouble, because the glagah *(Saccharum)* is not so high and „by far not so dense. Between the canes (i.e. *Saccharum*) a terres- „trial orchid with long reed-like stems and white flowers is fre- „quent". — This terrestrial orchid was *Arundina speciosa Bl.*, recorded already on p. 31 among the kremnophytes. Localities where such kremnophytes abound, are unfit for a luxuriant growth of *Saccharum*[1]). In the parts of Krakatao investigated by Penzig in 1897 this grass apparently was locally quite common, as may be concluded from 3 of the photo's of the interior, reproduced in his paper [2]). On each of these *Saccharum* is recognizable by its inflorescences.

The third grass named by Penzig, *Neyraudia*, is a robust strongly tillering, arundinaceous plant, reaching a height of 2—5 m. and deceivingly resembling a *Phragmites*. With the latter genus it has often been confounded, but besides by its villous glumae it can easily be distinguished by its massive culms. Old specimens often form clumps of some square meters' surface but I did never see in Java these clumps blend into a dense vegetation of very considerable extent. As to its habitat this species stands between the two former; it thrives quite well in a rather unfertile soil, it is often found in stony localities[3]), frequently even in great numbers though never covering large areas. But such unfavorable habitats as *Pennisetum* is contented with, — narrow, rather undeep crevices in rocks for instance —, seem, however, to be unfit for the growth of *Neyraudia*.

Penzig keeps silent about the individual frequency of the three species. He only says that between the grasses which formed in the valleys a nearly uninterrupted jungle, other plants sporadically occurred, so that the most suitable name for the vegetation in question was savanna or grass-steppe. In the rather humid parts with a deep soil *Saccharum* has probably predominated, forming there a rather dense, locally perhaps very dense vegetation, whilst in the rocky, less moist localities, besides a thin growth of low *Saccharum*-plants, *Neyraudia, Pennisetum* and other plants than grasses will have been found. Penzig says that, when he penetrated into the grass-jungles of the interior, numerous twining plants with their tough stems

---

[1]) But it should not be forgotten that, within a small space, considerable local differences may occur. Ravines may quite well bear a kremnophytic vegetation on the almost vertical parts of their walls and a *Saccharum*-jungle on the more inclined parts and near their bottom. One should also bear in mind that localities unfavorable for a luxuriant growth of *Saccharum*, but adjacent to favorable localities, may be constantly populated from the latter by means of the creeping rhizomes. In judging of a vegetation in the tropics one should always have a keen eye to its surroundings.

[2]) Ann. Jard. Bot. Buitenzorg XVIII (1902), pp. 101, 103, 104.

[3]) See also p. 53, footnote 2.

hampered his progress, but this statement should be received with some caution. Of the 5 species named by him in illustration of it, all of which are littoral plants, one is an *erect* shrub (*Scaevola frutescens Krause*), two are herbs with creeping but no twining stems (*Ipomoea pes caprae Sw.*[1]) and *Canavalia rosea D.C.*), the 2 others (*Vigna marina Merr.* and *Cassytha filiformis L.*) are indeed twiners but both are, like the 3 others practically restricted to the immediate neighbourhood of the sea, consequently to the outer border of the grass-jungle. None of these 5 species can at some altitude above sea-level have occurred in considerable numbers, though some stray specimen may have grown here or there.

When Penzig and his companions explored the steeper parts of the unweathered old basaltic slopes (from which the covering layer had been washed away very soon after the eruption), the aspect of the flora there appeared to be practically the same as in 1886. In other words, just as might be expected, the same kremnophytic vegetation was present that Treub had observed 11 years before [2]). Several species of ferns predominated in very many gregariously growing specimens, numberless prothallia were seen, in many places slimy blackish or bluish green *Schizophyceae* formed still the only covering of the rocks and the soil.

Of the components of the fern-association in 1897, Penzig, in his description of the vegetation of the rocky localities, mentions only 4, viz.: *Ceropteris calomelanos Und.* (Gymnogramme calomelanos Kaulf.), – *Acrostichum aureum L.*, *Stenochlaena palustris Bedd.* (Acrostichum scandens J. Sm.) and *Dryopteris setigera O. K.* (Aspidium vile Kze [3]). From the collective enumeration of these ferns as plants of the rocks clearly appears that Penzig did not pay to oecological factors the attention required for a scientific investigation. As a matter of fact these 4 species are never found together. *Ceropteris calomelanos* (See p. 54 under No. 1) and *Dryopteris setigera* (See p. 57 under No. 8) are halophobous kremnophytes; they must have been found on steep walls, in crevices of rocks or on ridges of pumice and ashes. The two other ferns are inhabitants of swampy or humid localities or at least of a soil with a high water-table. *Acrostichum aureum L.* prefers a saline substratum [4]) and occurs only exceptionnally

[1]) Very rarely the stems of *Ipomoea pes caprae* show some tendency to twine, but never they form a real obstacle.

[2]) One should bear in mind that Treub took part in this second botanical trip and could communicate his impressions to Penzig.

[3]) See p. 58, sub No. 8.

[4]) Cf. p. 55, No. 4.

in sweet water; *Stenochlaena palustris Bedd.* thrives as well where the water is very slightly brackish as where it is sweet [1]). If these 4 ferns really have been found near each other, the two first must have been absent in the locality where the 2 others occurred and conversely. One should always bear in mind that the bottom of a steep-sided ravine may be oecologically very different from its walls; the walls being frequently deficient in humus and very unfertile, consequently fit only for kremnophytes; the bottom more humid, often much richer in humus, hence as a rule much more fertile and bearing a quite different vegetation [2]). In vertical ravine-sides by some cause or other some parts may be dug out, the bottom of which not seldom becomes more or less horizontal and therefore less liable to erosion. This bottom, when not too young, usually bears a non-kremnophytic flora, quite different from that of the adjacent vertical parts. In the surroundings of Buitenzorg this phenomenon is very conspicuous.

After having ascended the rocks Penzig and his companions found several Phanerogams between grasses and in fern-associations. Here also the vegetation gave the general impression of a savanna. Grasses either small or large predominated; between them other herbs and shrubs occurred in scattered specimens. The vegetation varied but little, its components were few, a common occurrence with young vegetations in the tropics.

Two of the grasses of the somewhat higher parts of the mountain are recorded by Penzig under a specific name, viz. *Imperata cylindrica P.B.* (I. arundinacea Cyr.) and *Pogonatherum paniceum Hack.* (P. crinitum Trin). These two grasses also differ very much as to habitat. *Pogonatherum paniceum* is one of the commonest kremnophytes of Java: from the plains up to an altitude of 1700 m. it is spread throughout the island, except in regions where the east-monsoon, as a rule, is extremely long and dry. It grows exclusively on little weathered or much washed-off localities or in habitats oecologically identical with these: on steeps banks, on rocks and old walls, on big stones in rivers, frequently in great numbers. Banks of rivers and of hollow roads are often for a great length densely clothed with it; rarily it occurs in localities which are level or almost so. When the fruits are ripe, the then fragile axis of the inflorescence disarticulates into small fragments to each of which one fructiferous spikelet remains attached. These

1) See p. 54, No. 2.

2) See also C. A. Backer en Dr. F. van Slooten, Javaansche Theeonkruiden (1924), pp. 14 seq. On Krakatao the lower parts of the ravines were shortly after the eruption locally saliferous and consequently very fit to bear halophytes. Cf. p. 50, footnote 3.

spikelets are very small and provided with 2 thin, rather long awns; over land they are easily dispersed by wind. It is quite unknown whether animals also take part in their dispersal. The statement of Docters van Leeuwen [1]) that this species is spread by birds is apparently not based on actual observation, but seems to be a mere guess, lacking probability. Though in 1897 this grass may have plentifully occurred on steep banks on Krakatao, it cannot possibly have taken part in the savanna-vegetation proper.- *Imperata cylindrica P. B.* is the notorious alang alang, a grass spread all over Java from the plains up to an altitude of 2700 m. In favorable localities it multiplies with marvellous rapidity, not only by the growth and ramification of its rather deep-lying and far-creeping rhizomes but also by its fruits, which remain enclosed within the spikelets. These spikelets are clothed with very long white hairs; when ripe they disarticulate from their pedicels and are dispersed far and wide by the wind: one often sees them by thousands floating through the air. It is unknown how far the spikelets can be carried by the wind in a single flight but as to this matter I refer to what I have said on p. 101 on the dispersal of the fruits of *Saccharum spontaneum*.

Imperata is no kremnophyte; the plant never grows an vertical walls but on plains, ridges and slopes which, without being *very* dry, yet are not sufficiently humid for a dense growth of trees that would kill this heliophilous grass by their shadow; moreover it is very often found in deserted fields formerly used for agricultural purposes and on spots where forests have been destroyed by fire or by the axe. In such localities it often predominates in countless numbers, it forms those vast fields so vividly described by Junghuhn [2]). *Imperata* always grows intermixed with other plants which in a young alang-field may be hidden by the high grass-blades, but continually increase in size and number, wherever the alang-fields are left undisturbed. In the grass-jungle gradually young forest develops, except in too dry localities; the groves of trees at first form small islands which afterwards blend into each other [3]) and finish by killing out the grass.

[1]) Handelingen Eerste Natuurwetenschappelijk Congres (1920), p. 62.- Annales Jardin Botanique Buitenzorg XXXI (1921), p. 131.

[2]) Junghuhn, Java, Second Dutch Edition I, 289; German Edition I, 212.

[3]) The natives often wish to preserve the alang-fields which yield old leaves for thatching, young leaves for fodder. In order to renovate the pasture and destroy the young trees they set fire to the alang-alang in the east-monsoon, when many of the old leaves are dead and quite dry. The fire kills the young forest-growth but does not damage the deep-lying rhizomes of the grass which soon afterwards emit young leaves. Junghuhn [4]) wrongly meant that the alang-alang was set fire to in order to destroy it, just the reverse is aimed at. The fire is the pitiless foe of the young forest, the protecting friend of the grass.

[4]) Junghuhn, Java, Second Dutch Edition I, 296; German Edition I, 126.

P e n z i g states that in the investigated higher parts of the mountain *Imperata* was one of the commonest grasses, but he does not say in which kind of locality it occurred, neither whether it was already growing gregariously. Those parts of the ash- and pumice-fields which did not bear the much more robust and more moisture-needing *Saccharum spontaneum* may have been covered with a dense vegetation of alang-alang. On the rocks this grass can have grown only in deep crevices or must have been entirely absent.

In a locality of the somewhat higher parts of the mountain P e n z i g found many specimens of *Spathoglottis plicata Bl.* This is a violet- or white-flowered terrestrial kremnophytic orchid of sunny or slightly shadowed localities, by preference of such ones as are deficient in humus; it is spread throughout West- and Central-Java between 125 and 1600 m. altitude, especially above 600 m., except in regions where the east-monsoon is strong. It occurs principally on steep banks of ravines, terraces and hollow roads, moreover on rocks, in alang-fields, thickets and young secondary forests, locally often in rather great numbers. In 1908 this plant was frequent on vertical ravine-sides in the south-eastern part of Krakatao.

Together with *Spathoglottis,* but less frequently than this species, occurred in 1897 *Arundina speciosa Bl.,* likewise a terrestrial orchid, which is spread throughout Java from the plains up to 1600 m. except in regions where the east-monsoon is strong. It grows in the same kind of habitat as *Spathoglottis* and often intermixed with it. This species also may be locally rather frequent, especially in stony localities [1]) and in hard soils, provided the somewhat deeper layers of these, in which the plant roots, do not dry out too much in the east-monsoon.

P e n z i g says that these two orchids were *accompanied* by a third species, a *Phajus* of which he does not give the specific name and of which, to my knowledge, no specimens collected in 1897 on Krakatao are present in the Buitenzorg herbarium. The only species of this genus found until now on Krakatao is *Phajus tankervilliae*

---

[1]) In September 1897 it was found by Mr. A. E. K e r k h o v e n [See Teysmannia VIII (1898) p. 506] on the very stony lower slopes of Mount Guntur (20—40 rain-days in the 4 driest east-monsoon months) It was found back there i.a. by Dr. A. E r n s t in Nov. 1905 who in his description of the vegetation (Vegetationsbilder herausgegeben von Dr. G. K a r s t e n und Dr. S c h e n c k, 7e Reihe, Heft 1 and 2, IV) wrongly calls it a xerophyte. It roots in the deeper, moister layers of the soil which should not be confounded with the dry surface. See also the statement of F r a n s e n H e r d e r s c h e e cited on p. 102 herebefore

Bl. (Ph. blumei Lindl., Ph. incarvillei O.K.) which in May 1908 was found here and there on the south-eastern side of Krakatao in humid and not very dark ravines between 200 and 400 m. above sea-level. If — and such is highly probable -- it was this same species which was collected by Penzig in 1897, it did certainly not grow *in the company* of *Spathoglottis* and *Arundina*, i.e. in oecologically equivalent localities. For *Phajus tankervilliae Bl.*, a terrestrial orchid rather common in West-Java between 500 and 1200 m. above sea-level, is no kremnophyte but a mesophyte, it occurs on water-sides, in humid localities either stony or not and covered with grass or shrubs, less often in dense forests but never on the sterile walls, steep banks or sunny alang-fields preferred by the two other species mentioned by Penzig. The possibillity remains that *Phajus* was growing on the bottom of a ravine, *Spathoglottis* and *Arundina* in its vicinity on ravine-sides or on fields of ashes and pumice. The oecological difference between the bottom and the sides of a ravine was already discussed herebefore (p. 105).

It is worthy of note that from investigations made by Forbes [1]) appears that often autogamy takes place with the 3 species of orchids collected by Penzig [2]). There might be some connection between this fact and the abundance of *Spathoglottis* in 1897. Penzig gives no information about any insects having been found that might effect pollination.

Finally several specimens, often of more than one meter high, were found of *Blumea balsamifera D. C.* „und mehrere dieser verwandten Formen". For *Blumea balsamifera* see p. 51. One of the „allied forms" was probably *Erigeron linifolius*, as appears from a specimen preserved in the Buitenzorg Herbarium, wrongly recorded in Penzig's list under the name of *Blumea hieracifolia*. For this plant see p. 51. Neither *Blumea balsamifera* nor *Erigeron linifolius* lays high claims to the fertility of the soil, both *may* have been collected on arid ridges though they also occur in more fertile localities. Penzig says nothing definite about the habitat. Another „form" allied to *Blumea balsamifera* is *Pluchea indica Less.* (see p. 50), a species

---

1) Forbes, A Naturalist's Wanderings in the Eastern Archipelago (1885), p. 85 seq. and Knuth, Handbuch Blütenbiologie III² pp. 316, 318. 319.

2) With *Phajus tankervilliae* the same experience was made by J. J. Smith [see Die Orchideen von Java (1905), p. 196): „Bei allen von mir gesehenen Exemplaren fand „Selbstbefruchtung statt." — From an observation made by J. Krajenbrink [see De Tropische Natuur XVII (1928), p. 132 seq,] appears that, though autogamy takes place as a rule, there are some specimens that are non-autogamous. Smith has observed the latter particularity with specimens from Sumatra.

recorded in Penzig's list. This shrub only grows on the heavier soils either saline (and very frequently so) or not. Possibly it was collected in a locality with a badly pervious soil consisting of fine compressed ashes, such as was present in 1906 at the base of the rupture at Zwarte Hoek. It is very probable that part of the localities investigated by Penzig possessed a badly pervious soil; this may be deduced from the fact that many young specimens were found of *Ophioglossum reticulatum L.* (sensu lato), recorded by Penzig under the name of *O. moluccanum Schlecht.*) This is a terrestrial fern, occurring exclusively on badly pervious soils or on soils whose surface is covered by a badly pervious layer of fine mud, in not too dry localities, especially in the cultural zone, often in premises, on road-sides, in tea-plantations, rather frequently also in teak-forests. Penzig says nothing about the nature of the substratum on which he found this plant, neither does he tell which were the ferns forming associations (see p. 105) between the grasses in the higher localities reached by him. – Generally speaking the vegetation was rather monotonous and the number of species found was comparatively small.

Raciborski collected in 1897 on Krakatao on rocks and on the soil a number of Schizophyceae and Bacillariophyta. Of the first of these groups he gathered 6 species not found by Treub; the 6 species discovered by Treub in 1886 seem not to have been found back in 1897 [1]). This should not be considered proof that the old species had disappeared in 1897 and that the new ones were introduced after 1886; it is much more probable that in 1886 already numerous species of algae were present on the island and that both Treub and Raciborski collected only a few of these. The 2 species of mosses which Treub had collected in 1886 (see p. 60) were found back by Raciborski who collected furthermore a species of *Anthoceros* (Hepaticae), a species of *Hygrophorus* (Agaricaceae) and numerous specimens of the small Myxomyceta *Physarum cinereum Pers.* Here also Penzig does not give any further details.

After having ascended the mountain for about an hour, a time very much too short for allowing a somewhat intensive investigation, it became clear to Penzig and his companions that they could not possibly reach the top and they returned to their vessel. They cannot

1) Of course assuming that in both cases the determinations were correct, which cannot be decided as long as the specimens are not found back.

have reached a greater height than 200 à 300 m., it is even probable that they have remained below that. Consequently they explored only a very small part of Krakatao, the beach and the lower slopes on the north-western side, and the investigation was but a very cursory one. *Which was the vegetation of the other parts of the island in 1897 is fully unknown.* This must be taken into account when judging of the results of this trip.

The rocks and the pumice- and ash-fields explored by Treub and Penzig in 1886 and 1897 seem never afterwards to have been visited by a botanist, at least none of the subsequent authors on Krakatao has mede any communication about the vegetation on these slopes [1]). The investigations here came to an end with the excursion of Penzig. It is quite unknown whether since 1897 the aspect of the flora has been modified and if so in which manner. I myself think it probable that the basaltic rocks will continue to bear a thin vegetation of kremnophytes for yet many years to come. But I do not think an investigation there to be urgent: now that the starting-point of all investigations – viz. it being out of doubt that the vegetation was totally destroyed in 1883 – has disappeared, the kremnophytic vegetation on Krakatao cannot be considered to be more important than any other one and many of these kremnophytic vegetations may be reached and investigated much more easily and at much less cost.

After having left Krakatao Penzig and his companions paid a visit to Verlaten Eiland where they spent three hours in exploring the beach on the eastern side. From there they went to Lang Eiland, then since a short time inhabited by some officials of the Topographical Survey Service, where they in haste collected plants on the beach till it grew too dark. On the results of the trips to Verlaten Eiland and Lang Eiland I shall be silent; they are of no importance for the revegetation-problem of Krakatao.

Before mentioning the conclusions drawn by Penzig from his exploration I give a corrected list of all plants found by him on Krakatao, as far as possible with data about habitat and frequency. None of Penzig's plants is present in the Buitenzorg Herbarium, but there are some collected by Boerlage on this same trip.

---

1) Valeton (Cf. Chapter VII) tried in 1905 to climb them but had the misfortune to luxate an arm, by which accident he was compelled to give up the ascent.

Oscillatoriaceae.

1. Oscillatoria amphibia Ag.
2. Inactis vaginata Naeg. (Schizothrix vaginata Gom.)
3. Microcoleus chthonoplastes Thur.

Scytonemaceae.

4. Scytonema mirabile Bornem.

Chrooccaceae.

5. Gloeothece rupestris (Lyngb.) Bornem. var. *tepidariorum.*
6. Aphanothece castagnei (Breb.) Rabenh.

(Nos. 1–6:) In the interior, in many places clothing the rocks and the soil. Penzig does not give the frequency for each species separately.

Physaraceae.

7. Physarum cinereum Pers.
Habitat not mentioned. Occurred in many specimens.

Bacillariaceae.

8. Navicula spec.
9. Encyonema ventricosum (*Ag.*) *Grun.* (Plant of humid localities and of sweet, brackish or salt water).
10. Cystopleura gibba (*Ehrenb.*) *Kunze.*
11. „ *gibberula* var. *producta Grun.* Both species are plants of sweet and brackish water.
12. Nitzschia sigma (*Kütz*) *W. Sm.* Plant of brackish water.

(Nos. 8–12:) Habitat and frequency not mentioned.

Agaricaceae.

13. Hygrophorus spec.
Habitat and frequency not mentioned.

Anthocerotaceae.

14. Anthoceros spec.
Habitat and frequency not mentioned.

15, 16. The same two species of mosses as were in 1886 collected by Treub.

Polypodiaceae.

17. *Dryopteris unita* O.K. (Nephrodium cucullatum Baker). See p. 56, under No. 7.
Habitat and frequency not mentioned.

18. *Dryopteris setigera* O.K. (Aspidium vile Kunze). See p. 57, under No. 8.
According to Raciborski[1]) in 1897 very common on Krakatao.

19. *Nephrolepis exaltata* Schott. See p. 56, No. 6.
Habitat and frequency not mentioned.

20. *Blechnum orientale L.* See p. 55, No. 3.
Habitat and frequency not mentioned.

21. *Stenochlaena palustris Bedd.* (Acrostichum scandens J. Sm.). See p. 54, No. 2.
As habitat of this species Penzig mentions – but probably wrongly – steep rock-sides. According to Raciborski[2]) this plant occurred in 1897 in numerous specimens on Krakatao.

22. *Ceropteris calomelanos Und.* (Gymnogramme calomelanos Kaulf.) See p. 54, No. 1.
In 1897 well-developed specimens were found on steep rock-sides. Raciborski[3]) states that the plant was very common on the island in that year.

23. *Onychium siliculosum* C. *Chr.* (O auratum Kaulf.). See p. 60, No. 11.
Penzig says of this species that it was not found back in 1897. But according to Raciborski[4]), who was present at the same trip, this plant was then „sehr häufig auf der „Insel Krakatau". See p. 7.

24. *Pteris vittata L.* (recorded by Penzig under the name of Pt. longifolia L.). See p. 56, No. 5.
Habitat and frequency not mentioned.

25. *Pteridium aquilinum Kuhn* (Pteris aquilina L.). See p. 58, No. 9.
Habitat and frequency not mentioned.

---

1) Raciborski, Die Pteridophyten der Flora von Buitenzorg (1898), pp. 173, 174.
2) Raciborski, Die Pteridophyten der Flora von Buitenzorg (1898), pp. 54, 55.
3) Raciborski, Die Pteridophyten der Flora von Buitenzorg (1898), pp. 69, 70.
4) Raciborski, Die Pteridophyten der Flora von Buitenzorg (1898), pp. 163, 164.

26. *Drynaria quercifolia J. Sm.* (Polypodium quercifolium L.).

This is the well-known „oak-leaf fern" of the plains, which generally grows as an epiphyte on tree-trunks and heavy branches but sometimes also on rocks and steep banks. Closely allied to this species and united to it by intermediate forms is *Drynaria sparsisora Moore* (= Polypodium Linnaei Bory), which had perhaps better be regarded as a form of *Dr. quercifolia* and which, according to Raciborski [1], in 1897 occurred plentifully on Krakatao. A specimen collected there on the excursion of that year is present in the Buitenzorg Herbarium.

Habitat and frequency not mentioned by Penzig.

27. *Acrostichum aureum L.* See p. 55, No. 4.

As habitat of this species Penzig mentions — but without any doubt wrongly — steep rock-sides in the interior; he names it together with No. 22, *Ceropteris calomelanos Und.* Whoever has but a slight knowledge of the Javanese fern-flora knows that these two species — the first a halo-hygrophyte, the second a kremnophyte — never grow together. — Raciborski [2]) records *Acrostichum aureum L.* as occurring on Krakatao but says nothing about its habitat and frequency there.

Ophioglossaceae.

28. *Ophioglossum reticulatum L.* sensu lato (O. moluccanum Schlecht.). See p. 109.

This little fern reproduces itself as well by spores as by buds developing on the roots. Penzig and his companions found in 1897 numerous young specimens.

Lycopodiaceae.

29. *Lycopodium cernuum L.*

Club-moss spread throughout Java between 30 and 2200 m. above sea-level, except in regions where the east-monsoon is strong. It grows in sunny or moderately shadowed localities and is a typical kremnophyte growing by preference on unfertile soils, very often on the steep sides of hollow roads, terraces, ravines and of the small dams which in Java so very often border the roads. This species found in 1897 and

---

1) Raciborski, Die Pteridophyten der Flora von Buitenzorg (1898), p. 118.
2) Raciborski, Die Pteridophyten der Flora von Buitenzorg (1898), pp. 50, 51.

later years on Krakatao many fit habitats; it has been collected there on almost every excursion.

Penzig is silent on the habitat and frequency of this species in 1897.

Pandanaceae.

30. *Pandanus tectorius Sol.* (P. littoralis Jungh.). See p. 43, No. 4.
This species was numerous on the beach. Penzig found its fruits in the drift-zone at Zwarte Hoek. In 1886 Treub had found in this same drift-zone *Pandanus*-fruits, probably belonging (partly at least) to the same species.

Gramineae.

31. *Imperata cylindrica P. B.* (I. arundinacea Cyr.). See p. 106.
Of this plant Penzig only says that it was one of the commonest grasses in the higher localities reached by the excursionists.

32. *Saccharum spontaneum L.* See p. 100.
With *Pennisetum* (No. 35) and *Neyraudia* (No. 37) this grass formed locally a dense vegetation on the hills and in the valleys. Probably in that vegetation *Saccharum* greatly predominated.

33. *Pogonatherum paniceum Hack.* (P. crinitum Trin.). See p. 105.
Penzig states that in the higher localities reached by the excursionists this plant was one of the commonest grasses. Nothing is said about its habitat.

34. *Ischaemum muticum L.* See p. 89.
In great numbers on the beach at Zwarte Hoek.

35. *Pennisetum macrostachyum Brongn.* (Gymnothrix elegans Büse). See pp. 48 and 53.
This grass is said to have formed, together with *Saccharum* (No. 32) and *Neyraudia* (No. 37), a dense vegetation on the hills and in the valleys. It is very improbable that it was one of the chief components of this vegetation.

36. *Spinifex littoreus Merr.* (Sp. squarrosus L.). See p. 89.
In great numbers on the beach of Zwarte Hoek.

37. *Neyraudia madagascariensis Hook. fil.* var. *zollingeri Hook. fil.* See p. 103.

This plant is said to have formed, together with *Saccharum* (No. 32) and *Pennisetum* (No. 35), a dense vegetation on the hills and in the valleys. It is quite possible that here and there groups of this grass were growing but it is not very probable that it was one of the chief components of the vegetation.

Cyperaceae.

38. *Cyperus pennatus Lamk.* (Wrongly named by Penzig C. digitatus Roxb.).

39. *Fimbristylis spathacea Roth.*

See for these two species p. 47, No. 5 and 6; p. 90. Both species were numerous on the beach of Zwarte Hoek.

Orchidaceae.

40. *Arundina speciosa Bl.* See p. 107.

In the higher localities reached by Penzig this species was growing in the company of *Spathoglottis plicata Bl.* (No. 42), but was less numerous than this. Nothing is told on the nature of substratum.

41. *Phajus tankervilliae Bl.* See p. 107.

Probably this was the plant mentioned by Penzig under the name of *Phajus* spec. According to Penzig it occurred in the company of *Spathoglottis plicata Bl.*, but was less numerous. On p. 108 I argued already that it is highly improbable that these two species were growing on the same substratum. On the nature of the habitat Penzig is silent.

42. *Spathoglottis plicata Bl.* See p. 107.

In the higher localities reached by Penzig this species was numerous. Nothing is told on the nature of the habitat.

Casuarinaceae.

43. *Casuarina equisetifolia Forst.* See p. 91, No. 6.

Habitat and frequency unknown.

Moraceae.

44. *Ficus toxicaria L. fil.*

Small tree, rather common in West-Java between 300 and 1300 m. above sea-level. It occurs in sunny or moderately shadowed localities especially on not too steep ravine-sides, along rivers and in young secondary forests; locally it is often very common. The reproduction is effected exclusively by the figs which are sessile by pairs in the axils of the leaves.

Penzig says nothing about habitat and frequency of this species, nor does he tell whether the trees found bore already figs and whether these figs contained gall-wasps.

Lauraceae.

45. *Cassytha filiformis L.* See p. 91, No. 7.

Penzig says nothing about habitat and frequency of this species.

Leguminosae.

46. *Albizzia chinensis Merr.* (A. stipularis Boiv.)

Tree, spread throughout Java from the plains up to ± 1300 m. above sea-level. It grows in sunny or slightly shadowed localities, especially on periodically very dry soils, in grassy plains, in thickets, in secondary forests, very frequently also along water-courses, locally often in great numbers. The reproduction is exclusively effected by the seeds. The thin pods do not burst when ripe but break into fragments, each of which contains one or more seeds; these fragments are easily conveyed by water and over some small distance also by wind.

About habitat and frequency of this tree Penzig tells nothing. In the Buitenzorg Herbarium there are no materials present collected on this excursion on Krakatao; the tree seems never to have been found back there, if it was found at all. I think it by no means impossible that the specimen in question was wrongly named; it might have been *A. retusa Bth.*, a tree of sandy beaches that in 1908 was found on the beach of Zwarte Hoek.

47. *Canavalia rosea D. C.* (In Penzig's paper under the name of *C. obtusifolia D. C.*). See pp. 89 and 92.

According to Penzig this creeper — the only species of the genus found on Krakatao — was numerous on the beach of Zwarte Hoek.

48. *Vigna marina Merr.* (In Penzig's paper mentioned as *V. lutea D. C.* and *V. luteola D. C.*). See pp. 89 and 93. Numerous on the beach at Zwarte Hoek.

Euphorbiaceae.

49. *Euphorbia atoto Forst.* See pp. 90 and 93.
Numerous on the beach at Zwarte Hoek.

Malvaceae.

50. *Hibiscus tiliaceus L.* See p. 93.
Penzig says nothing about habitat and frequency of this species.

Sterculiaceae.

51. *Melochia umbellata Stapf* (M. indica A. Gray).
Small, fast-growing, heliophilous, halophobous tree which lays but few claims to the fertility of the soil and thrives as well in constantly rather humid (but not marshy) soils as in periodically very dry ones. It is spread throughout Java from the plains up to an altitude of ± 1700 m., especially beneath 1500 m. and often grows in deserted fields invaded by *Imperata,* where it is one of the pioneers of the young secondary forest that is to cover them in time. Moreover it is found in grass-jungles, thickets and in light forests, locally often in numerous but usually scattered specimens, which already from afar catch the eye by their whitish crown. The reproduction is exclusively effected by the light winged seeds.
Penzig says nothing about habitat and frequency of this species.

Lecythidaceae.

52. *Barringtonia asiatica Kurz* (B. speciosa Forst.). See p. 43, No 5.
Penzig says nothing about habitat and frequency of this species. In the drift-zone at Zwarte Hoek he found its fruits washed ashore.

Combretaceae.

53. *Terminalia catappa L.* See p. 39, No. 2.
Penzig says nothing about habitat and frequency of this species. In the drift-zone at Zwarte Hoek he found its fruits washed ashore.

Apocynaceae.

54. *Cerbera manghas L.* (C. odollam Gaertn.) See p. 46, No. 3.
Penzig says nothing about habitat and frequency of this species. In the drift-zone at Zwarte Hoek he found its fruits washed ashore.

Convolvulaceae.

55. *Ipomoea pes caprae Sw.* See p. 48, No. 7.
This creeper was numerous on the beach at Zwarte Hoek.

Rubiaceae.

56. *Morinda citrifolia L.* See p. 93, No. 16.
Penzig tells nothing about habitat and frequency of this species.

Goodeniaceae.

57. *Scaevola frutescens Krause* (Sc. Koenigii Vahl.). See p. 49, No. 9.
This plant was numerous on the beach at Zwarte Hoek. Penzig found its fruits also in the drift-zone.

Compositae.

58. *Vernonia cinerea Less.*
Herb spread throughout Java from the plains up to an altitude of 1400 m. It grows in sunny or slightly shadowed, not too humid localities, especially on grassy road-sides and in lawns, moreover as a rather common weed in gardens and in fields; it is also rather numerous in many teak-forests. The reproduction is effected by the small fruits which are produced in large numbers and possess a well-developed pappus.
Penzig says nothing about habitat and frequency of this species.

59. *Erigeron linifolius Willd.* See p. 51.
Perhaps this plant was one of the forms allied to *Blumea balsamifera* which Penzig found in the higher localities reached. He makes no further communication about its habitat and frequency.

60. *Blumea balsamifera D. C.* See p. 51.

Penzig found several species of more than 1 m. height of this undershrub (which reaches in Java a height of 4 m.) in the higher localities reached. For the rest he gives no details about habitat and frequency.

61. *Pluchea indica Less.* See pp. 50 and 108.

Possibly this shrub was another of the forms allied to *Blumea balsamifera* which Penzig found in the higher localities. I think it, however, more probable, that he found it in the older parts of the beach at the base of the rupture, where it was also found in 1906. Penzig says nothing about its habitat and frequency.

62. *Senecio sonchifolius Moench* (Emilia sonchifolia Moench). See p. 52.

Habitat and frequency unknown.

63. *Wedelia biflora D. C.* (In Penzig's paper named Wedelia asperrima Boerl., W. scabriuscula Boerl. and Wollastonia spec.). See p. 49 under Wollastonia.

Penzig says nothing about habitat and frequency of this species.

This list contains 16 lower Cryptogams, 13 vascular Cryptogams and 34 Phanerogams, in total 63 species. Treub's list gives 8 lower Cryptogams, 11 vascular Cryptogams and 15 Phanerogams, in total 34 species. Penzig, for reasons unknown to me not computing the lower Cryptogams, comes for the three islands of the Krakatao-group together to a total of 62 species (50 Phanerogams and 12 vascular Cryptogams) whilst Treub in 1886 found only 15 Phanerogams and 11 vascular Cryptogams. Penzig asserts not to have found back seven of Treub's species, viz. three Phanerogams (*Hernandia sonora, Wollastonia sp.* and *Tournefortia argentea*) and four ferns (*Nephrodium calcaratum Hook.,– Nephrodium flaccidum Hook.,– Pteris marginata Bory* and *Onychium auratum Kaulf.*). But he thinks it possible that these species were overlooked. Assuming that they still occurred in 1897 Penzig arrives for the 3 islands at a total of 53 Phanerogams and 16 ferns. The 26 „littoral species" whose seeds or fruits he found washed ashore (Cf. pp. 96, 97) he leaves out of account, „as it is not quite certain – though highly probable – that they will „come to development on the Krakatao-group".

This résumé should be corrected in this manner, that the *Wollas-*

tonia spec. of Treub is the same species as *Wedelia biflora D.C.*, hence, as a matter of fact, was found back; that the plants named by Treub *Nephrodium calcaratum* and *Nephrodium flaccidum* were respectively *Dryopteris unita O.K.*, and *Dryopteris setigera O.K.*, both of which were found back by Penzig, whilst finally also *Onychium siliculosum C. Chr.* (O. auratum Kaulf.) was found back in 1897 on Krakatao. As to Phanerogams, besides *Hernandia* and *Tournefortia*, the species of *Erythrina* and *Calophyllum inophyllum L.* collected in 1886 by Treub were in 1897 not found back on the main island. Assuming with Penzig that all species found by Treub still occurred in 1897 on Krakatao we come for this island itself to a total of 22 lower Cryptogams, 14 vascular Cryptogams and 38 Phanerogams, in all 74 species. Of the 26 „littoral species" whose seeds or fruits had been found washed ashore, only 22 were indicated by a specific name. Of these no more than 13 are littoral plants (See p. 97).

Penzig also takes for granted that the entire flora of the Krakatao-group was introduced after 1883. As agents he considers sea-currents, wind and birds. Introduction by man, according to him, has not taken place, as the islands till 1897 were quite uninhabited and were but rarely visited and only for a short time. Some pages after this statement he admits however the possibility that man could have accidentally introduced a couple of species of *Ficus* [*F. hispida L. fil.* and *F. septica Burm.* (= F. leucantatoma Poir.)] into Lang Eiland which had been inhabited during a short time and had received more visitors. The vascular Cryptogams were, according to him, exclusively introduced by the wind. For the 53 Phanerogams of the entire group (38 on Krakatao itself) he ascribes introduction by wind to 17 species (14 on Krakatao itself) possessing small fruits or seeds, hence 32 % (for Krakatao itself 37 %) of the total number of species. The number of species whose fruits or seeds were carried over by sea-currents is very much higher. Penzig records as such 32 species (for Krakatao itself 25 species) or 60 % of the total number. For the greater part these are — as Penzig rightly observed — typical littoral plants. Of only 4 of the species collected by him Penzig supposes that the seeds were carried to the Krakatao group by frugivorous birds or bats. Only one of these (3 % of the total number), *F. toxicaria L. fil.*, was collected on Krakatao.

When one compares the list of the finds of Treub with that of Penzig who examined much the same locality, both lists prove to be very much alike as regards ferns. Of the 11 fern-species of 1886

10 were found back in 1897; Penzig and his companions collected only 3 ferns or fern-allies not found by Treub. If one assumes that all these ferns were introduced from outside the island after the eruption it is singular that within the 3 years elapsed between the eruption and Treub's visit no less than 11 ferns were introduced in the examined locality; contrariwise but 3 ferns or fern-allies in the 11 years elapsed between Treub's visit and that of Penzig. If, on the contrary, we assume that most or all of the ferns found by Treub had inhabited before the eruption the basaltic slopes or the higher ravines (see pp. 66—69) and that a number of specimens had escaped destruction by means of spores or rhizomes, consequently that but very few species had to be introduced form outside the island, the extraordinary concordance between the fern-flora of 1886 and that of 1897 is quite naturally explained.

For the first time were in 1897 found 3 species, viz. *Drynaria quercifolia,* a fern which usually lives on trees but frequently also on rocks or on steep walls, *Ophioglossum reticulatum L.,* a rather common fern of badly pervious or very stony soils and the club-moss *Lycopodium cernuum L.,* one of the commonest kremnophytes of Java. These 3 species also may long before the eruption have lived in unfertile localities or on steep walls near Zwarte Hoek and may have survived the catastrophe. Treub's investigation in 1886 of the very accidented locality must necessarily have been very cursory and several species may have been present that he did not see. Penzig himself admits that even in 1897, when 5 botanists and a very able native plant-collector examined the locality, species may have been overlooked. Moreover, one should always bear in mind that in 1886 as well as in 1897 *only a small part of the island was examined* and this only hastily; consequently that when a species was not *found* this by no means furnishes proof that it did not *occur*. Therefore we have no right to take for granted that the 3 species not found by Treub in 1886 were introduced after that year. It is not even certain that they were *introduced* at all after the eruption; they may have *survived* it. Of course I will not deny the *possibility* that after 1883 a number of plants were introduced. For the halophytes and the typical littoral plants this is next to certain, as the original beach of Krakatao partly was destroyed by the eruption, partly was permanently and deeply buried, and it is not probable that the interior, before the eruption, offered any fit habitat to halophytes or typical littoral plants. These are the following: *Acrostichum aureum L.,*—

Pandanus tectorius Sol.,— Ischaemum muticum L.,— Spinifex littoreus Merr.,— Casuarina equisetifolia Forst.,— Cassytha filiformis L.,— Hernandia peltata Meissn.,— Erythrina variegata L. var. *orientalis Merr.* [1]),— *Canavalia rosea D.C.,— Vigna marina Merr.,— Euphorbia atoto Forst.,— Hibiscus tiliaceus L.,— Calophyllum inophyllum L.,— Barringtonia asiatica Kurz,— Terminalia catappa L.* (also on riversides),— *Cerbera manghas L.,— Ipomoea pes caprae Sw.,— Tournefortia argentea L. fil.,— Morinda citrifolia L.,— Scaevola frutescens Krause* and *Wedelia biflora D.C.* Together 21 species, among which 20 Phanerogams. The spores, seeds or fruits of all these are fit to be spread by sea-currents and the seeds or fruits of the Phanerogams are often found in the drift-zone. Hence one may with some right suppose that these species were carried over by the sea, though it is of course possible that animals and man have also taken some part. *Cyperus pennatus Lamk.* and *Fimbristylis spathacea Roth,* though neither obligatory halophytes nor true littoral plants, are often found close to the sea in localities either saline (and frequently so) or not. Penzig found them in great numbers on the beach at Zwarte Hoek; their fruits may quite well have been carried over by sea-currents.

The 16 Phanerogams of the interior found by Penzig are:

1. Imperata cylindrica P.B.,— 2. Saccharum spontaneum L.,— 3. Pogonatherum paniceum Hack.,— 4. Pennisetum macrostachyum Brongn.,— 5. Neyraudia madagascariensis fil. var. *zollingeri Hook. fil.,— 6. Arundina speciosa Bl.,— 7. Phajus tankervilliae Bl.,— 8. Spathoglottis plicata Bl.,— 9. Vernonia cinerea Less.,— 10. Erigeron linifolius Willd.,— 11. Blumea balsamifera D.C.,— 12. Pluchea indica Less.,— 13. Senecio sonchifolius Moench,— 14. Ficus toxicaria L. fil.,— 15. Albizzia chinensis Merr.,— 16. Melochia umbellata Stapf.* Penzig considers all these plants as introduced ones, No. 1—13 by the wind, No. 14 by animals, No. 15 and 16 by the sea. From his publication does not appear, however, that this opinion is based on any experiment or actual observation; its only base is apparently the fact that the first 13 species possess very light or winged or pappose fruits or seeds, that the fruits of No. 14 are included within a fleshy receptale, whilst No. 15 has light fruits which may be transported by water. Penzig apparently did not know that also the seeds of No. 16, *Melochia umbellata,* are light and winged. If he had known,

[1]) Whether in 1886 this species was found by Treub or *E. fusca Lour.* which as a rule inhabits river-sides and sweet-water swamps is uncertain.

he would have followed the old recipe and have taken the wind for the agent of dispersal.

I have already repeatedly argued that it is by no means proven that the original vegetation of Krakatao was entirely destroyed by the eruption but that many species may have survived the catastrophe. Of the 16 species recorded above, No. 1, 2, 3, 4, 5, 6, 8, 9, 10, 11, 12, 13 and 16 may quite well have grown long before the eruption on the slopes of Zwarte Hoek. No. 7, 14 and 15 may be remnants of the former forest. We know, however, *nothing* with *certainty*, neither that species have been spared, nor that they have been introduced. The Krakatao-problem has no base, there is no fixed point of depart, one can only guess and always guess, nothing has been properly investigated, nothing has been proven, nor *could* be proven by the defective methods of investigation which were followed.

It is, of course, quite possible that seeds or fruits of plants now living in the interior were introduced after the eruption. But even if such an introduction might have taken place, the fact that 14 of the plants named above possess light or winged or pappose seeds or fruits is by no means proof that these seeds or fruits were carried over by the wind. On p. 11 I have already argued that such seeds and fruits often get into rivers which can spread them further. Hence for these species also transport by sea-currents may not be considered impossible, as long as it has not been proven that the short time they have to remain in the sea-water – the transport by sea-currents from Java to Krakatao can be made within a single day – suffices to destroy the germinative power. Then remains the possibility of transport by pumice, driftwood, animals and man. Especially the fruits of *Compositae* and *Gramineae* often adhere to garments or goods, hence may be easily spread by man.

As regards the seeds which have already reached Krakatao or which are produced there, the wind certainly takes and has taken an active part in their distribution over the island, especially before the vegetation was very dense, though certainly distribution by rain-water flowing down also took place on a large scale, as will be admitted by everyone who has witnessed tropical showers in the mountains. Nothing is known of the fauna of Krakatao in the first ten years after the catastrophe. But we may safely assume that the superterranean fauna was entirely, or almost so, driven away or destroyed by the eruption and it may be that the number of flower-pollinating insects and seed-dispersing animals in the first years after the eruption was small. This would explain that in the first years in the interior especially such higher plants rapidly multiplied whose flowers are pollinated by the wind (the 5 grasses) or are frequently autogamous (the 3 orchids and the 5 Compositae) and which moreover possess seeds or fruits which may be easily spread over land by the wind or by rain-water flowing down. On this point there exists however no certainty *for we know absolutely nothing of the vegetation of the higher parts of the mountain in those years.*

No. 15, *Albizzia chinensis,* is a tree which in the plains often inhabits river-sides. Its pods or the fragments of these possess floating power; therefore this species, if it has really been found, might, as Penzig supposes, have been carried to Krakatao by sea-currents.

Remains *Ficus toxicaria,* which, in spite of its ominous name, is not toxical at all. The figs of many Javanese species of *Ficus* are sought after by birds and frugivorous bats. It is certain that in Java *Ficus hispida* and *F. septica* are spread by such bats; they often spring up under the margin of penthouses which serve to these animals as a resting-place and beneath which one very frequently finds the dropped remnants of figs carried over by them from elsewhere. Possibly the big figs of *Ficus toxicaria* also are eaten by birds or bats. But this may not be considered proof that these animals have *introduced the species into the island,* as long as it has not been observed, that they, after having eaten the fruit on any other island, fly over to Krakatao and drop the seeds there. On the very accidented island, many parts of which are most difficultly accessible and whose flora always was but very imperfectly known, such observations could hardly be made. But as long as they have *not* been made, the supposition that in this manner plants have been introduced by animals, *remains a mere guess, nothing more, and certainly no proof for such an introduction across a broad sea-arm.*

The results of the excursion of 1897 are summarized by Penzig in 3 conclusions:

1. *The islands which in 1883 were completely deprived of their vegetation, receive but very slowly a new flora. The total number of the vascular plants observed is not yet trebled in* $10^1/_2$ *years (on Krakatao in June 1886 26 species, in March 1897 52* [1]*), hence double the number of species; 62 on the entire group).*

[1]) Including the 5 species collected in 1886 by Treub but not found back in 1897. See p. 119, 120.

2. *On all three islands of the Krakatao-group the beach bears mainly a pes-caprae-formation. Mangrove is entirely absent. Only on Lang Eiland forest has begun to develop. More in the interior the vegetation has the character of a savanna or a grass-steppe. This vegetation consists partly of grasses of more than a man's height which form a dense jungle in localities fit for them. On the hills and the ridges less tall grasses, numerous ferns and scattered Phanerogams are found. On the rock-sides, now like 10 years before, ferns greatly predominate. Shrubs are few in number, trees are almost entirely absent. It will take a long time before the higher layers of the soil are sufficiently weathered and possess a sufficiently quantity of humus for bearing again a dense forest.*

3. *Most of the Phanerogams were introduced by sea-currents, fewer by the wind, very few by animals.*

These conclusions cannot possibly be agreed with. Penzig in 1897 — like Treub in 1886 — investigated but a very small part of the island. Nothing is known of the flora of the localities not investigated in those years; hence from the fact that in 1897 species were found that were not seen during the most incomplete investigation of 1886, may by no means be deduced that these arrived after 1886. Consequently *nothing* can be said with certainty about the increase of the number of species between 1886 and 1897. For neither of these years the total number then present can be even guessed at with some approach to accuracy.

It is certain that in 1897 the beach at Zwarte Hoek bore a greater number of species and a much denser vegetation than in 1886, but all littoral plants found there may have occurred in 1886 elsewhere on the beach of Krakatao. It is quite unknown when they first reached the island. Whether plants of the interior had already been introduced in 1897 and, if so, which plants, from where, when and in which manner is utterly unknown. Littoral forest was in 1897 seen only on Lang Eiland but may quite well have been present on non-investigated parts of the beach of Krakatao. By far the greater part of the beach remained unexamined in that year, and this part was not even visible from Zwarte Hoek. It is quite conceivable that on Krakatao no mangrove-vegetation was found, for the coasts of the island are unfit for such a vegetation. But some stray specimens of species, as a rule growing in the mangrove, may have occurred here and there on the beach [1]).

---

[1]) Cf. p. 98, footnote 1.

Wherever the investigated localities had been covered in 1883 with a *thick and permanent* layer of ashes and pumice they doubtless received a new flora after that year, but it is unknown whether this flora originated from outside the island or from the washed-off parts where the original soil had come to the surface again. It is certain that the secondary vegetation in the investigated parts was thin in 1886, locally much denser in 1897. Quite analogous phenomena may be observed in not too dry regions of Java. There also the secondary vegetation begins with a thin vegetation of grasses soon becoming denser and often intermixed with ferns; on this point the investigation of Krakatao has disclosed nothing that was not already known long before from elsewhere. But the *real* problem: „has indeed the new flora of Krakatao entirely been intro- „duced from elsewhere and, if so, in wich manner?", has remained completely unsolved even unattacked by the explorations of Treub and Penzig. It is quite conceivable that in steep rocky localities deficient in humus the kremnophytic vegetation predominated in 1897 as it did in 1886; probably that vegetation will maintain its character for years.

It is very *improbable* that the lack of forest in the investigated parts of Krakatao was exclusively or mainly due to unfertility of the soil. It does not appear that Penzig has instituted any investigation on this point. On the mere strength of Treub's assertion of 1886 he has taken the unfertility for a proven fact. Soils fit for a luxuriant *Saccharum*-vegetation are in a tropical region receiving so much rain as Krakatao sufficiently fertile for the development of forest. But, if forest is to develop, two conditions have to be fulfilled, viz. sufficient import of seeds fit to germinate and enough time for them to develop into trees. Hence we have to ask: Had the investigated localities long enough before 1897 received an amount of seeds sufficient to permit the development of forest.

In my opinion this was not the case. Though I fully admit the possibility that seeds of inland plants in some way or other were imported from elsewhere, I believe that this import did not taken place on a large scale or, at least, that not many of the seeds imported found a locality fit to germinate in and to develop further. The revegetation of the interior of the island had in my opinion in the first years to originate mainly from seeds or spores *produced on the island itself.* In the higher ravines on the east-, south-east- and south-side, where the original soil soon was laid bare, the forest probably soon regenerated from its remnants, but many years had to pass before these young trees could produce so many fruits that a rich forest-vegetation all over the island could arise. Possibly lack of fruit-dispersing animals has also played a part but this can no more be examined now. The fields of pumice and ashes near Zwarte Hoek that were investigated by Penzig are situated in such a locality that seeds from the higher ravines on the south-east- and east-side of the island cannot at all be carried there by rainwater and hardly by wind. A too scanty import of seeds therefore can be the main cause of the lack of forest in localities fit for it.

It is quite well possible that in 1897 the higher part of the ravines on the south-south-east and east-side of Krakatao bore already young forest. One should remember that Treub in 1886 from the sea, hence at a great distance, saw already plants near the top of the mountain, and that therefore these plants must have been rather large. It is quite possible that they were shrubs or trees.

As to the mode of import, it is next to certain that the whole of the littoral vegetation was new and had been mainly or exclusively introduced by the sea. It is utterly unknown from where the plants found in the interior came, nor how they were imported and in which numbers.

If we now proceed to state the reliable results given by Penzig's trip we can formulate them as follows:

1. The littoral vegetation at Zwarte Hoek was in 1897 much richer and denser than in 1886. The oldest parts of the beach (that were at best 14 years old) had in 1897 already been leached out and bore locally a rich growth of the halophobous *Saccharum spontaneum*. The copious rain-fall and the situation of the beach at the base of a high steep wall are the cause of that leaching-out having taken place so rapidly.

2. As might be expected, the kremnophytic vegetation of the steep walls of the interior has remained practically the same between 1886 and 1897.

3. The vegetation of the investigated parts of the interior had grown much denser and had in 1897 much the same aspect as young secondary vegetations in Java in rather moist climates. It consisted mainly of grasses and ferns with scattered Dicotyledons. Those plants formed in the moister localities a tall and dense vegetation, on the less moist ridges a thinner and less tall one.

Especially such plants were numerous as are able to rapidly multiply vegetatively and can do without animals for the pollination of their flowers and the dispersal of their seeds or fruits.

4. It is fully unknown whether between 1886 and 1897 plants have been introduced and, if so, in which manner.
5. Nothing is known of the fauna of 1897 and its relations to the vegetation.
6. In 1897 no examination has been instituted after the degree of weathering and the humidity of the soil. But from the fact that *Saccharum* occurred gregariously on ash- and pumice-fields may be safely deduced that these localities were neither lacking in fertility nor in humidity and were by no means unfit for the development of forest.
7. Hence the lack of forest in the examined parts cannot be ascribed to edaphic or climatic conditions but must have another cause, probably a quite unsufficient import of seeds from other localities either on the mountain or outside it.
8. Nothing was in 1897 brought to light about the composition and origin of the vegetation of localities which were most important for the investigation, viz. the higher ravines.

VII. THE TRIP OF GOLENKIN AND VALETON TO KRAKATAO IN 1905.

On March 1st 1905 a trip to Krakatao was made by the Russian savant Golenkin, mainly for the purpose of collecting marine algae. Dr. Th. Valeton, the author of the systematic part of the „Bijdragen tot de kennis der Boomsoorten van Java [1])" took — merely as a tourist — part in this trip which was never intended to be a botanical exploration of the island. On this occasion he collected some plants with the assistance of Arsin († 1913), a native functionary of the Buitenzorg Herbarium. Valeton has never published anything on the botanical results of his trip, but kindly sent me by letter some information about a list [2]), made by Arsin of the plants collected in 1905, to which list Dr. Valeton added another one recording the plants merely observed but not collected by him. The above mentioned list of Arsin was inserted by Docters van Leeuwen (sub No. 3) in his lists of plants found on the islands of the Krakatao group [3]).

From the letter of Valeton it appears that he landed first on the north-west side of the island near Zwarte Hoek. After having investigated the flora of the beach he went some way up the steep basaltic slopes where he had the misfortune to luxate an arm so that he had to give up further ascent. The locality investigated bore a dense vegetation of glagah (*Saccharum spontaneum L.* [4]). Afterwards the visitors landed on the east side of the island. On this side, where the parts of the interior adjacent to the beach are much less steep than at Zwarte Hoek, the explorers penetrated some little way into the jungle, but their harvest was poor. Some of the collected plants were found back by me in the Buitenzorg Herbarium. Part of them bore the addition Krakatao ilir, another part Krakatao oedik, which additions were made by Arsin. Valeton kindly informed me about the meaning

[1]) This most excellent and useful work was written in collaboration with Dr. S. H. Koorders († 1919) who had collected the materials but took no part in the systematical elaboration. He added the small-type details about distribution, leaf-shedding, velocity of growth, reproduction, cultivation, properties and use of the wood, vernacular names a. s. o. The systematic part of volume 12 was written by Dr. J. J. Smith, that of volume 13 partly by Dr. Valeton, partly by Dr. J. J. Smith.

[2]) Dr. Valeton kindly gave me this list which is now in my possession.

[3]) Handelingen Eerste Nederlandsch-Indisch Natuurwetenschappelijk Congres, Weltevreden (1920) pp. 54 seq.— Ann. Jard. Bot. Buitenzorg XXXI (1921) pp. 126 seq.

[4]) See pp. 100, seq.

of these additions: by Krakatao ilir Arsin indicated the locality at Zwarte Hoek, by Krakatao oedik the locality on the eastern side.

This trip is remarkable because then for the first time the eastern side of Krakatao was visited by botanists. But only the very lowest parts of the mountain, which had been deeply and permanently buried by the eruption of 1883 could be explored and this only hastily, the higher slopes being everywhere secluded from the beach by a very dense girdle of *Saccharum spontaneum*, which could not possibly be crossed during so short a trip. None of the plants found back by me in the Buitenzorg Herbarium was accompanied by any data about frequency and habitat, barring the additions mentioned above.

In the list beneath the plants mentioned by Valeton and Arsin are recorded in systematic order. Of the plants found back by me in the Buitenzorg Herbarium I have mentioned on which side of the island they were collected. Of the rest it is unknown where they were found.

Polypodiaceae.

1. *Dryopteris unita O.K.* See p. 56 under No. 7.
2. *Nephrolepis exaltata Schott.* See p. 56, No. 6.
   Zwarte Hoek.
3. *Ceropteris calomelanos Und.* See p. 54, No. 1.
4. *Onychium siliculosum C. Chr.* See p. 60, No. 11.
5. *Drynaria quercifolia J. Sm.* See p. 113, No. 26.

Lycopodiaceae.

6. *Lycopodium cernuum L.* See p. 113, No. 29.

Cycadaceae.

7. *Cycas rumphii Miq.*

Dioecious tree with a robust trunk, up to 2 m. high and 2–8 dm. thick [1]), under cultivation [2]) growing very slowly

1) Cf. Hasskarl in Tijdschr. Ned.-Indië IV, p. 129.— in Flora (Regensb. Bot. Zeitung) 1847, p. 490 and Junghuhn, Java, 2nd Dutch Edition I, 268 in ann.— German Ed. I, 195 in ann.***

2) This *Cycas* is in Java very often cultivated for ornamental purposes, mostly in ♀ specimens as especially the big cones of ♂ during the anthesis spread a strong and very disagreeable odour; the smell of flowering ♀ specimens is less strong. These male cones are much frequented by pollen-collecting small bees. The artificial reproduction is most frequently made by taking off the lateral shoots which sprout from old trunks. Cultivated specimens, as a rule, produce no ripe seeds. This may be caused by the comparatively rare occurrence of ♂ specimens in cultivation, so that flowering ♀ and ♂ specimens are but seldom found in each other's vicinity. Where both sexes are growing intermixed in rather large numbers and in a wild state, as is the case here and there on the southern coast of Java, ripe seeds are frequently found.

and flowering at rather long intervals. This species, which is often confounded with C. *circinalis L.* [1]), is spread over the entire southern part of Java from the beach up to 50 m. above sea-level; moreover it has been found in a few localities near the north-coast and on some coral islands in the Java-sea. It often, though by no means exclusively, grows quite near the sea, as well in the Barringtonia-formation as in thickets and forests somewhat back from it; as already Junghuhn [2]) rightly observed, it may be found at some kilometers' distance from the shore. In some regions (i. a. in the south of the residency of Kediri [3]), along the southern margin of the Gunung Sèwu [4]) and at the Sand-bay) it is rather numerous, but as a rule it occurs in scattered specimens. The dry seeds can remain floating on sea-water during a long time; Ernst found them in the drift-zone of Krakatao [5]). Probably they are also spread by animals; if not, the occurrence of the tree high above sea-level in forests in the interior could hardly have a natural cause. Schimper [6]) found on Nusa Kambangan seeds, the fleshy outer layer of which seemed to have been gnawed away by hermit-crabs.

„Somewhere on the beach" Valeton found a specimen of about 2 meters height, thus an old plant. Probably this specimen was found on the eastern side. If a *Cycas* of that dimensions had still occurred a year afterwards at Zwarte Hoek, Ernst and his companions who then explored this locality with much care would certainly have observed it. It is not impossible that the specimen seen by Valeton was the same as that observed by the explorers of 1906, the trunk of which was 1,65 m. high (See Chapter VIII). Valeton's plant was not found back by me in the Buitenzorg Herbarium, the *Cycas*-materials of that institute in 1924 having already been lent away since many years.

---

1) The latter species has not yet been found in the Malayan Archipelago.

2) Junghuhn Java, 2nd Dutch Edition I, 268, 316,— German Edition I, 195, 231, under the name of *Cycas circinalis*. Cf. also Blume in Rumphia IV (1848), p. 14 (under the name of *Cycas circinalis*): „in vicinia litorum, praesertim in saxosis „calcareis centum pedes supra maris aequor".

3) Teysmann in Natuurk. Tijdschr. Ned. Indië XI, 137 (under the name of *Cycas circinalis*).

4) Cf. Junghuhn, Reisen, pp. 108, 112 (under the name of C. *revoluta*).

5) Ernst, Neue Flora Vulkaninsel Krakatau (1907), p. 29.

6) Schimper, Indo-Malayische Strandflora (1891), pp. 75 and 175, tab. VII. fig. 10.

Gramineae.

8. *Saccharum spontaneum L.* See pp. 100 seq.
Numerous at Zwarte Hoek. In the letter mentioned above Valeton speaks of „endless glagah". But he may have got that impression as well on the eastern side.

9. *Ischaemum muticum L.* See p. 89.
Zwarte Hoek.

10. *Spinifex littoreus Merr.* See p. 89.

Cyperaceae.

11. *Cyperus pennatus Lamk.* See p. 47, No. 5; p. 90.
Zwarte Hoek.

12. *Remirea maritima Aubl.*

A few data about this herb [1]) were already given on p. 47. The spikelets are united into a kind of head at the top of the culm. The small fruits when falling off remain enclosed within a few glumes, of which especially the topmost one consists of a thick corky floating tissue. – The plant is spread from West- to East-Java; it is one of the first immigrants establishing on young beach-walls.

On the east-side of the island.

Orchidaceae.

13. *Spathoglottis plicata Bl.* See p. 107.

Casuarinaceae.

14. *Casuarina equisetifolia Forst.* See p. 91, No. 6.
Zwarte Hoek.

Ulmaceae.

15. *Trema orientale Bl.* (inclusive of *Tr. amboinense Bl.*).
Fast growing tree, sometimes bearing nitrogen-tubercles on its roots, spread throughout Java from the plains up to an altitude of 2400 m. It thrives as well in jungles and secondary forests as in primeval woods, as well on poor soils as on rich ones and is – especially in the mountainous regions – one of the first species appearing in fields and

---

1) A reproduction of a drawing is printed in De Tropische Natuur VIII (1919), p. 9, fig. 15.

tea-plantations which have been deserted on account of the soil being exhausted. It also occurs – and frequently in large numbers [1]) – on young lahars (mudstreams rapidly descending or having descended from a volcano). It is reproduced by its small drupes which are sought after i.a. by birds.

Zwarte Hoek.

Moraceae.

16. Ficus fistulosa Reinw.

Small tree with either axillary, rameal or cauline figs, spread throughout Java from the plains up to an altitude of 2100 m., especially above 500 m. The tree cannot withstand a long drought; hence it mainly occurs in regions where the east-monsoon is feeble; elsewhere it is restricted to watersides and depressions of the soil. It requires no fertile soil and likes a not too dark habitat; it grows by preference on forest-borders, on ravine-sides, in grass-jungles, in thickets and young secondary forests, in which it may be one of the most common species. It is also found in light forests (teak-forests, exceptionally also in Casuarina-forests), but in dark primeval forests it is absent or almost so.

Zwarte Hoek.

17. Ficus fulva Reinw.

Small tree with axillary figs [2]), spread throughout Java but almost exclusively in regions where the east-monsoon

[1]) This species is very often confounded with *Parasponia parviflora Miq.*, which deceivingly resembles it and bears the same vernacular names *(Anggring, Anggrung, Kuraj)*, but may be easily distinguished by the connate stipules and the broadly imbricate segments of the perianth. – As a synonym of *Parasponia* Miquel (Plant. Jungh. p. 69) and Junghuhn (Java, 2nd Dutch Edition I, 549; German Edition I, 400) cite *Celtis montana Jungh.*, but an authentical specimen of this last species preserved in the Buitenzorg Herbarium is *Trema orientale Bl.* [Cf. Smith in Koorders and Valeton, Bijdrage No. 12 tot de Kennis der Boomsoorten van Java (1909) p. 664]. Like Smith I think it quite likely that Junghuhn has confounded the 2 species. Hence it is possible that the gregariously growing Anggring found by Junghuhn in Sept. 1846 on the lahars of Gunung Kelut [See Java, 2nd Dutch Edition I, 550; III, 679, 692; German Edition I, 400; II, 471, 480), and named by him *Parasponia parviflora* was, as a matter of fact, *Trema orientale*. On the same volcano Dr. Ph. van Harreveld saw in 1921 this *Trema* growing gregariously on a lahar which had flown down but two years before. To the same species may belong the tree mentioned by Junghuhn in the second Dutch Edition of Java (II, 158) as *Celtis*, in the German edition (II, 116) as *Parasponia*, which he saw in August 1837 in immense numbers on the hills and ridges of the gigantic lahar which had flown from M. Galunggung in Oct. 1822.

[2]) By some mistake Ernst (Neue Flora Vulkaninsel Krakatao (1907) p. 31) records this species among the trees which „ihre Blüten und Früchte gleich zahlreichen anderen „Tropenbäumen nicht an den jüngsten Zweigen sondern am Stamm und den älteren „Ästen bilden". Of the 6 species of *Ficus* found in or before 1906 on the islands of the Krakatao-group only *F. hispida* is always cauliflorous and *F. fistulosa* often so.

is feeble; elsewhere it is restricted to watersides. It prefers a not too dark habitat and occurs mainly on river- and ravine-sides, on forest-borders and especially in young secondary forests; in dark primeval forests it is lacking.

From a note of Dr. Valeton it appears that numerous rather young specimens of this tree were found, but it is not clear on which side of the island.

18. *Ficus hirta Vahl.*

Erect shrub, found in Java only in the western half between 15 and 600 m. above sea-level, as well in regions where the east-monsoon is rather strong as where it is feeble. It occurs in sunny or slightly shadowed localities in grassy wilds, in jungles, in alang-fields and young secondary forests.

Zwarte Hoek.

Urticaceae.

19. *Pipturus incanus Wedd.*

Small dioecious tree, spread throughout Java from the plains up to ± 1200 m. above sea-level, especially in regions where the east-monsoon is feeble. Where the east-monsoon is strong the tree is restricted to watersides or to localities with a high watertable. It prefers a not too dark habitat, it is found in grassy wilds and in jungles, in secondary forests, on forest-borders and ravine-sides, in light parts of primeval forests. The reproduction is effected exclusively by fruits which, included in the thickened fleshy tops of the axes of the inflorescence, form with them white spurious berries.

Lauraceae.

20. *Cassytha filiformis L.* See p. 91, No. 7.

Leguminosae.

21. *Derris heterophylla (Willd) Backer* (= D. uliginosa Bth).

Creeping or twining shrub of the borders and the lighter parts of the mangrove, mainly occurring in swampy localities and along or near river-mouths, often in numerous specimens. On sandy beaches the plant is much more rare. The reproduction is effected by the indehiscent pods, which are provided with a thick light rind, hence possess floating power.

On the eastern side.

22. *Erythrina variegata L.* var. *orientalis Merr.* See p. 46, No. 1. Young specimens (as appears from a note of Dr. Valeton). It does not appear where these were found.

23. *Mucuna acuminata D. C.*
Twining shrub, spread throughout Java from the plains up to 1400 m. above sea-level, on forest-borders, in thickets and light forests, often also on river-sides, which in Java are frequently bordered by thickets and trees. The reproduction is effected by the seeds which seem sometimes to possess floating power, sometimes not [1]. As a rule the pods burst before falling off. Ripe pods which by some cause have fallen before bursting, possess floating power, because of the fruit-cavity not being entirely filled up by the seeds.
Zwarte Hoek.

24. *Canavalia rosea D. C.* See pp. 89 and 92, No. 8.
Zwarte Hoek.

25. *Vigna marina Merr.* See pp. 89 and 93, No. 9.
On the eastern side.

Euphorbiaceae.

26. *Macaranga tanarius M. A.*
Small tree, spread throughout Java from the plains up to ± 1400 m. above sea-level, in thickets, secondary forests and the lighter parts of primeval forests. The tree is not particular as regards the fertility of the soil, it thrives as well on arid calcareous rocks and in other very stony localities as in fertile ravines and in village groves. The reproduction is effected by the fleshy fruits which are eaten by birds, i.a. by *Calornis strigatus Horsf.* (= C. chalybea Horsf.). The seeds found in the intestines of a bird of this species shot on Krakatao were sown by Docters van Leeuwen in the Buitenzorg Botanical Garden and grew up into trees [2]. For the value of this experiment see p. 71.
On the eastern side.

[1] Possibly this depends on the degree of desiccation. I saw some fresh ripe seeds of this species sink at once in sweet water but Schimper [Indo-Malayische Strandflora (1891), p. 164] states that seeds of *Mucuna* possess floating power because of the seed-cavity not being entirely filled up by the embryo.

[2] Dr. W. Docters van Leeuwen, The Vegetation of the Island of Sebesy, situated in the Sunda-Strait, near the Islands of the Krakatao-group; in the year 1921, in Ann. Jard. Buitenzorg XXXII (1923), 160.

27. *Euphorbia atoto Forst.* See pp. 90 and 93, No. 10.
On the eastern side.

Vitaceae.

28. *Columella trifolia Merr.* (= Vitis trifolia L).
Shrub, high climbing with the aid of tendrils, spread throughout Java from the plains up to an altitude of 600 m. It grows in hedges, thickets and not too dark forests, often in large numbers and occurs in two forms, a glabrous one (*Cayratia carnosa Gagnep.* —, *Cissus carnosa Lamk.*), which seems to grow only near the sea, and a pubescent form (*Cissus cinerea Lamk.*) which grows from near the sea to far in the interior. It is unknown whether in this case there exists a correlation between edaphic factors and the presence or absence of pubescence. The reproduction is effected, besides by shoots sprouting from far-creeping horizontal roots (which in Java make the plant a troublesome weed in sugar-cane fields), by the shining black berries which as they cause a troublesome irritation of the throat cannot be eaten by man. Whether they are sought after by animals is unknown to me, but one often sees plants loaded with hundreds of ripe berries from which fact one would be inclined to deduce that, generally spoken, the berries are not much relished. The berries possess floating power, the small hard seeds not.
Zwarte Hoek (the glabrous form).

29. *Leea aequata L.*
Erect shrub, spread throughout Java from the plains up to 800 m. above sea-level, exceptionally even higher. It grows in thickets, in young secondary forests and in teak-forests. The reproduction is effected by the berries. Plants damaged by fire easily emit new shoots from the crown of the root [1]).
On the eastern side.

Malvaceae.

30. *Hibiscus tiliaceus L.* See p. 93, No. 11.
Zwarte Hoek.

---

1) J. G. B. Beumée, Floristisch-analytische onderzoekingen van de korte flora in kunstmatig aangelegde djati-plantsoenen op Java, in verband met de ontwikkeling van den djati-opstand (1922), p. 90.

31. *Thespesia populnea Soland.*

Small tree, spread throughout Java on sandy beaches, especially in the Barringtonia-formation. The reproduction is effected by the capsules whose pericarp bursts into 5 valves, thereby separating from the tough fibrous indehiscent endocarp which encloses several seeds. This endocarp possesses floating power because the fruit-cavity is not entirely filled up by the seeds; the seeds themselves also can float on sea-water [1]).

Young specimens (as appears from an annotation made by Valeton).

Guttiferae.

32. *Calophyllum inophyllum L.* See p. 44, No. 6.

On the eastern side.

Lecythidaceae.

33. *Barringtonia asiatica Kurz.* See p. 43, No. 5.

34. *Barringtonia insignis Miq.*

Tree of dark constantly humid forests, spread throughout Java from the plains up to 600 m. above sea-level. It is rather numerous in the south-western part of the Residency of Bantam (Java) which is adjacent to the Sunda Straits. The reproduction is effected by the fruits, on the mode of spreading of which nothing is known.

The tree never afterwards having been found on the island the correctness of the determination is somewhat liable to doubt. But it cannot be controlled, materials being lacking.

Convolvulaceae.

35. *Ipomoea pes caprae Sw.* See p. 48, No. 7.

36. *Ipomoea spec.*

Verbenaceae.

37. *Premna integrifolia L.*

Erect shrub spread throughout Java on sandy or clayey (provided not-swampy) beaches, moreover in many localities in the interior to many kilometers from the sea. The reproduction is exclusively effected by the juicy shining black drupes possessing each a 1–4-seeded pyrene. These pyrenes

1) See Schimper, Indo-Malayische Strandflora, pp. 159, 163.

possess floating power [1]), hence may be spread by water. But it is probable that animals also take some part in distributing the plant; if not, its occurrence in many places of the interior far from water could not be explained.

Labiatae.

38. *Hyptis brevipes Poit.*

Erect annual of American origin introduced into Java already very long ago and at present spread all over the island, but especially in its western and central parts, from the plains up to an altitude of 900 m. It grows in fields, on the narrow dikelets (galangan) between rice-fields and along water-courses, locally often in great numbers. The reproduction is effected by the small nutlets which possess floating power.

Zwarte Hoek.

Rubiaceae.

39. *Uncaria sp.*

Genus of scandent shrubs with small winged seeds. Several species occur in Java.

40. *Nauclea sp.*

Genus of trees and shrubs with small winged seeds, of which 5 species occur in Java.

Young specimens (according to Valeton). Possibly the species found by Valeton was *Nauclea purpurascens Korth.*, which in 1908 and afterwards was found in the island.

41. *Guettarda speciosa L.*

Erect shrub or small tree, occurring throughout Java on not-swampy beaches, especially on sand- and coral-coasts, usually in scattered specimens. The reproduction is effected by the rather large drupes which are hardly fleshy but possess a thick light mesocarp [2]) and hence floating power. Guppy collected them in the drift-zone of the Keeling islands and found that they can remain floating for months [3]).

---

1) See Guppy, Observations of a Naturalist in the Pacific between 1896 and 1899, II (1906), p. 560 seq.

2) See also Engler und Prantl, Natürl. Pflanzenfam., IV⁴, p. 97. fig. C, D.

3) Guppy, Observations of a Naturalist in the Pacific between 1896 and 1899, II (1906), p. 529.

On the eastern side.

42. *Morinda citrifolia L.* See p. 93, No, 16.

Cucurbitaceae.

43. *Trichosanthes bracteata Voigt.*

Shrub, high climbing with the aid of tendrils, with bright red many-seeded fruits which possess a fleshy yellow mesocarp and a dark green, slimy endocarp. It is spread throughout Java from the plains up to 1500 m. above sea-level, as well in regions where the east-monsoon is rather strong as where it is feeble. It grows in thickets, on forest-borders and in light woods, generally in scattered specimens. The reproduction is effected by seeds, about the mode of spreading of which nothing is known.

Compositae.

44. *Erigeron linifolius Willd.* See p. 51.
Zwarte Hoek.

45. *Blumea balsamifera D.C.* See p. 51.

46. *Erechthites hieracifolia Rafin.* See p. 52.

47. *Pluchea indica Less.* See pp. 50 and 108.
Zwarte Hoek.

48. *Wedelia biflora D.C.* See p. 49, No. 1.
Zwarte Hoek.

From a comparison of this list with that of the finds of Penzig (Cf. pp. 111 seq.) it appears that Penzig and his companions found in 1897 at Zwarte Hoek 47 Phanerogams and higher Cryptogams 18 of which were littoral or halophilous species, whilst Valeton collected there 18 plants among which were 6 littoral or halophilous species or, assuming for the moment that all plants of unknown habitat were found at Zwarte Hoek, 39 species of which 12 littoral or halophilous ones. It would be absurd to conclude from this fact that in the 8 years passed between the two excursions the vegetation of the locality had grown much poorer. This appears at once from the following survey:

		1897	1905	1906
1.	*Pandanus tectorius Sol.*	numerous	not mentioned	many large specimens
2.	*Pogonatherum paniceum Hack.*	one of the commonest grasses	not mentioned	very numerous
3.	*Pennisetum macrostachyum Brongn.*	many specimens	not mentioned	numerous
4.	*Fimbristylis spathacea Roth.*	numerous	not mentioned	several specimens
5.	*Scaevola frutescens Krause.*	numerous	not mentioned	several large specimens
6.	*Vernonia cinerea Less.*	present	not mentioned	several specimens

It appears from this survey that the above 6 species were present both in 1897 and in 1906, 5 of them in numerous specimens. It is not to be believed that they were lacking in 1905, the less if one takes into account that 4 of them are perennials and that two (No. 1 and No. 5), both large shrubs or small trees, were represented in 1906 by adult specimens. A much more plausible explanation of them not having been recorded by Valeton is that the investigation of 1905 was less accurate or less extensive than those of 1897 and 1906. And the only conclusion that can be safely drawn from this list is that conclusions as to a vegetation growing poorer of richer can never safely be drawn from very incomplete lists.

The eastern side, which Valeton was the first to investigate, yielded 10 species (of which 7 littoral or halophilous ones) or, assuming that all plants of unknown habitat were found on this side of the island, 31 species of which 13 littoral or halophilous plants. The next year Ernst and his companions found on the south-east and east side 51 plants, 36 of which were littoral or halophilous species. From this may only be concluded once more that the investigation of 1905 was less extensive or less accurate than that of 1906. It could hardly be otherwise, the one-day tourist-trip of Valeton was never meant to be a scientific excursion nor was it executed as such. A complete and careful investigation of the difficultly accessible locality would have required many months.

The following plants were in 1905 found for the first time on Krakatao:

1. *Cycas rumphii Miq.*
2. *Remirea maritima Aubl.*
3. *Trema orientale Bl.*
4. *Ficus fistulosa Reinw.*
5. *Ficus fulva Reinw.*
6. *Ficus hirta Vahl.*
7. *Pipturus incanus Wedd.*
8. *Derris heterophylla Backer*
9. *Mucuna acuminata Grah.*
10. *Macaranga tanarius M. A.*
11. *Columella trifolia Merr.*
12. *Leea aequata L.*
13. *Thespesia populnea Soland.*
14. *Barringtonia insignis Miq.*
15. *Premna integrifolia L.*
16. *Hyptis brevipes Poit.*
17. *Uncaria sp.*
18. *Nauclea sp.*
19. *Guettarda speciosa L.*
20. *Trichosanthes bracteata Voigt.*

I shall revert to this list in my discussion of Ernst's paper (Chapter VIII).

VIII. THE EXPEDITION OF ERNST, PULLE, CAMPBELL AND BACKER TO KRAKATAO IN APRIL 1906.

[Dr. A. Ernst, Die Neue Flora der Vulkaninsel Krakatao in Vierteljahrsschrift der Naturforschenden Gesellschaft in Zürich LII (1907) Heft 3].

On April 26th 1906, 9 years after the trip of Penzig and his companions, one year after that of Golenkin and Valeton, the islands Krakatao and Verlaten Eiland were visited by Ernst, Pulle, Campbell and myself [1]). The results of this trip were published by Ernst in the clever and well-written paper mentioned above, by far the best that ever appeared on the revegetation of Krakatao [2]), though this gifted author was, like all his companions, severely handicapped by lack of knowledge of the tropical flora and by the fact that not yet fully one day could be given to the exploration of the much accidented and densely overgrown locality, a time altogether unsufficient for an accurate botanical investigation. The principal merit of Ernst's publication is that he has brought together a great number of facts and speculations which are of considerable value for the *study* of the problem of revegetation though they have contributed little or nothing to its *solution*. It might be compared to a finger-post indicating the way to an interesting spot but not going there itself.

The investigators, who were coming back from a visit to the south-western point of Java, reached Krakatao from the south-east. Leaving out of account the vague statement of Haeckel in 1901 (See p. 27, footnote 3) and the very few finds of Valeton (See p. 130 seq.) this was the first time, 23 years after the eruption, that attention was paid to the vegetation of the eastern and south eastern side of the island, where as I argued on p. 26, part of the old vegetation may have survived the catastrophe, especially in the higher ravines, which therefore ought to have been carefully explored as soon as possible after the eruption.

---

1) See also Jaarboek van het Departement van Landbouw in Nederlandsch Indië (Annuary of the Department of Agriculture in the Dutch East Indies) 1906, pp. 4, 5. Instead of „zuidwest" and „westen" (south-west and west) one should read there „zuidoost and oosten" (south east and east),

2) By some curious oversight neither this important publication nor even those of Treub and Penzig have been mentioned in the list of litterature on Krakatao contained in Overzicht der Nederlandsch-Indische Natuurmonumenten (Survey of the Nature Monuments in the Dutch Indies) composed by Dr. K W. Dammerman, published in 1924 by the Dutch Indian Society for the Preservation of Nature. This list gives of botanical litterature only the papers of Docters van Leeuwen, discussed in Chapter XI, hereafter.

The first impression made on him and his companions by the vegetation of the south eastern side is vividly described by Ernst [1]) in these words: „Mit steigender Verwunderung gewahren wir bei der „Annäherung an die Ostküste von Krakatao die erstaunlichen Fort- „schritte der Vegetation. Fast die ganze Südostseite vom Strande „bis an den Gipfel und den Rand des steilen Abhanges ist mit Grün „überdeckt. An der Südostküste, wo wir zuerst zu landen gedenken, „läuft dem Strande parallel ein Waldgürtel in welchem sich aus der „Ferne schon die zahlreichen graugrünen Casuarinen erkennen lassen. „Weiter südlich erheben sich neben schlanken Laubbäumen mit quirlig „gestellten Ästen [2]) die dunkeln Wedel einiger Kokospalmen. Auch „auf der langsam zum Fusse des Kegelberges ansteigenden Ebene „sind vereinzelte Bäume und Sträucher sichtbar. In einigen Schluchten „auf halber Höhe des Berges vereinigen sie sich wieder zu waldar- „tigen Beständen, und weitere isolierte Bäume und Sträucher sind auch „an den obersten Abhängen und auf dem Gipfel zu erkennen".

This description of the main features of the vegetation in that year is in substance correct; I only have to add that the forest did not occur in a few but in several ravines especially on the southern side. We were indeed greatly astonished at the progress which, in our opinion, the vegetation had made in the 9 years elapsed since 1897. None of us considered that all [3]) former explorations had taken place on the opposite, the north western side of the island and that nothing at all was known of the flora of the south eastern side before 1906, so that the vegetation found by us might not be compared with that of the other side in 1897. Another serious shortcoming was that all of us—like everyone then—on the mere authority of Treub took the total destruction of the vegetation by the eruption of 1883 for a well-proven fact, not liable to any doubt. Often since I have severely blamed myself for that unpardonable levity.

Two parts of the island were visited by us. First we landed on the south-eastern side where we examined the drift-zone [4]) and the vegetation of the beach, which we followed for a rather long way in a northern direction. From the south eastern angle we tried to reach the top of the mountain or at least the higher ravines, but

[1]) Ernst, Neue Flora Vulkaninsel Krakatau (1907), p. 28.

[2]) *Terminalia catappa L.*

[3]) Except the trip of Valeton in 1905. But of that trip the results had not yet been published.

[4]) We found i.a. seeds or fruits of *Barringtonia asiatica Kurz,—Calophyllum inophyllum L.—Cerbera manghas L.,—Cycas rumphii Miq.,—Nipa fruticans Wurmb,—Pandanus tectorius Sol.,—Terminalia catappa L., Xylocarpus* spec. and several other species, the names of which were not noted down.

the strongly accidented locality and the dense *Saccharum*-vegetation presented so many obstacles, that notwithstanding our utmost exertions, we were unable to effect our purpose [1]) in the short time we could dispose of; we did not reach a greater height than at best 40 or 50 m., perhaps even less. Then we went back to the shore, embarked again, rounded the eastern and the northern coast and landed at Zwarte Hoek, where we investigated the same beach as Treub had done in 1886, Penzig in 1897, Valeton in 1905. The interior of the island above Zwarte Hoek, of which our predecessors had likewise examined the flora, was not explored in 1906; only on the very base of the basaltic rocks some plants were collected.

The plants collected in 1906 are recorded beneath in 4 lists, of which the first contains the finds made at Zwarte Hoek, the second the vegetation of the beach on the south-eastern side, the third the vegetation of the examined part of the interior on the south eastern side, whilst the fourth contains some plants mentioned by Ernst for this trip but not observed by me. I thought it advisable not to follow the example given by former authors on Krakatao who united all their finds into a single list without indicating the habitats, because in this way it becomes impossible to compare the vegetation of different localities.

A. FINDS AT ZWARTE HOEK.

The locality, described already on page 36, consists mainly of a level beach extending from Zwarte Hoek in a south-eastern direction. This beach is ± $^3/_4$ km. long; it consists of sand, ashes and pumice and bears on the land-side several big boulders. In 1897 it was probably still narrow [2]); if so, it has since that year increased in size; in 1908 it was in its broadest part ± 150 m. [3]) wide; in 1906 it can have been but little narrower. Only the younger parts of this locality were in 1906 a beach in the oecological sense of the word, their vegetation consisted mainly of halophytes; the older parts, especially those close to the base of the rupture, were leached out by the frequent rains and possibly also by ground-water moving in a horizontal direction. From the fact that *Saccharum spontaneum*

---

1) See Ernst, Neue Flora Vulkaninsel Krakatau (1907), p. 32–34.

2) Cf. Penzig in Ann. Jard. Bot. Buitenzorg XVIII (1902), p. 97. Treub, who visited the same locality in 1886, says nothing on the breadth of the beach in that year.

3) According to the map in Jaarverslag van den Topographischen Dienst in Nederlandsch-Indië (Year-book of the Topographical Survey Service in the Dutch Indies) 1908, plate XXIII.

was growing luxuriantly and gregariously [1]) at the base of the rupture may be deduced that the soil there was practically salt-free. This locality, consisting of two oecologically very different parts, is secluded from the interior of the island by a steep wall at whose base fallen ashes have formed here and there small tali. On the south-eastern side the beach ends in a very large talus of ashes, which in 1906 was still unscalable and devoid of vegetation; on the north-western side it ends in the black basaltic rocks from which Zwarte Hoek (i.e. Black Corner) has derived its name. The lowest parts of these rocks bore in 1906 a meagre vegetation.

In 1906 46 species were collected at Zwarte Hoek. These were:

Polypodiaceae.

1. *Dryopteris unita O. K.* (Nephrodium cucullatum Baker). See p. 56, sub No. *7*.

A few specimens on the base of the steep wall formed by the rupture.

2. *Nephrolepis exaltata Schott.* See p. 56, No. 6.

Several rather large groups on the older, leached-out parts of the littoral plain.

3. *Ceropteris calomelanos Und.* (Gymnogramme calomelanos Kaulf.). See p. 54, No. 1.

Numerous on the base of the rupture.

4. *Drynaria quercifolia J. Sm.* (Polypodium quercifolium L.). See p. 113, No. 26.

A few specimens of this usually epiphytic fern were growing in crevices at the base of the basaltic rocks.

Lycopodiaceae.

5. *Lycopodium cernuum L.* See p. 113, No. 29.

A few feeble specimens at the base of the steep wall formed by the rupture.

Pandanaceae.

6. *Pandanus tectorius Sol.* (P. littoralis Jungh.). See p. 43, sub No. 4.

Several specimens as well on the younger parts of the beach as on the older parts, many of them of a large size.

[1]) See herebefore pp. 94 and 100 seq. and the very instructive plate in Ernst, Neue Flora Vulkaninsel Krakatau (1907), tab. IV, fig. 7.

Gramineae.

7. *Imperata cylindrica P.B.* (I. arundinacea Cyr.). See pp. 106, 114.
Some specimens in crevices at the base of the basaltic rocks.

8. *Saccharum spontaneum L.* See pp. 100 seq.
Very numerous on the leached-out older parts of the littoral plain along the base of the steep wall formed by the rupture.

9. *Pogonatherum paniceum Hack.* (P. crinitum Trin.). See p. 105.
Numerous on the steep wall formed by the rupture.

10. *Ischaemum muticum L.* See pp. 89 and 90.
Numerous on the beach.

11. *Pennisetum macrostachyum Brongn.* (Gymnothrix elegans Büse; Pennisetum elegans Ernst). See p. 48, No. 8 and p. 99.
A few specimens in crevices at the base of the basaltic rocks; numerous on the older leached-out parts of the littoral plain.

12. *Spinifex littoreus Merr.* (Sp. squarrosus L.) See pp. 89 and 90.
Scattered on the younger parts of the beach.

Cyperaceae.

13. *Cyperus pennatus Lmk.* (In Ernst's paper under the names of Cyperus digitatus Roxb. and Mariscus umbellatus Vahl.) See p. 47, No. 5; p. 91, No. 4 and p. 90.
Scattered on the older, firmer parts of the beach.

14. *Fimbristylis spathacea Roth.* See p. 47, No. 6; p. 90 and p. 91, No. 5.
Several specimens on the older, firmer parts of the beach.

Palmae.

15. Cocos *nucifera L.* See p. 39, No. 3.
Some young, possibly planted specimens behind the flood-mark. The coco-nuts washed ashore were all empty.

Casuarinaceae.

16. *Casuarina equisetifolia Forst.* See p. 91, No. 6.
A few scattered young specimens. These cannot have been the same as those found by Penzig in 1897 which, if they had remained in life, would in 1906 have grown up into tall trees. Such trees we did not see at Zwarte Hoek.

Ulmaceae.

17. Trema orientale Bl. (Inclusive of *Tr. amboinense Bl.* under which name this species is mentioned in Ernst's paper). See p. 132, No. 15.

Several specimens, among which some rather large ones, on the older parts of the beach.

Moraceae.

18. Ficus fulva Reinw. See p. 133, No. 17.

Several young specimens on the older parts of the littoral plain.

Lauraceae.

19. Cassytha filiformis L. See p. 91, No. 7.

Many specimens, especially on the older parts of the beach, parasitizing on diverse plants.

Leguminosae.

20. Pithecolobium umbellatum Bth.

Small spinous tree, spread from West- to East-Java but almost exclusively on heavy soils near the sea, often on the inner border of the mangrove, less frequently on sandy or rocky coasts, exceptionally in the interior. The reproduction is effected exclusively by the articulate indehiscent pods which, when ripe, easily break into fragments. These fragments possess floating power because of their rind being thick and light; the seeds sink in water.

One specimen on the older firmer part of the littoral plain.

21. Sophora tomentosa L.

Erect shrub, restricted to beaches rich of lime and deficient in humus. It is spread from West- to East-Java and chiefly occurs on coral coasts but also on coasts that are rich of shells. The reproduction is effected exclusively by seeds; the pods, which after having reached full maturity remain attached to the plant for yet a long time, do not burst, but their rind slowly decays till at last the seeds fall out. These seeds possess floating power in consequence of the cotyledons consisting of a spongy tissue [1]).

A few specimens on the younger parts of the beach somewhat behind the flood-mark.

---

[1]) Schimper, Indo-Malayische Strandflora (1891), p. 165.

22. *Desmodium umbellatum D. C.*

Erect shrub, spread from West- till East-Java, especially on sandy beaches, exceptionally on river-sides. The reproduction is effected by the small articulate pods, which, when ripe, break into fragments which possess floating power because of their rind being thick and light.

A few specimens on the younger parts of the beach.

23. *Derris heterophylla Backer* (D. uliginosa Bth.). See p. 134, No. 21.

A few specimens on the younger parts of the beach.

24. *Erythrina variegata L.* var. *orientalis Merr.* (E. indica Lmk.). See p. 46 sub No. 1.

A few young specimens on the younger parts of the beach.

25. *Canavalia rosea D. C.* (In Ernst's paper named C. obtusifolia D. C.) See p. 92, No. 8.

Numerous on the younger parts of the beach.

26. *Vigna marina Merr.* (V. lutea D. C. and V. luteola D. C.). See pp. 89 and 93.

Numerous, especially on the younger parts of the beach.

Euphorbiaceae.

27. *Euphorbia atoto Forst.* See p. 90 and p. 93, No. 10.

Several specimens on the younger parts of the beach.

Sapindaceae.

28. *Dodonaea viscosa Jacq.*

Erect shrub or small tree, the reproduction of which is effected exclusively by seeds. When ripe, the 2–3-winged 2–3-locular capsules either fall off as a whole or burst open. Each cell contains 1–2 seeds which are provided with a black, moderately thick testa and remain for some time attached to the placenta after the valves have fallen off. The distribution takes place either over land by the wind which drives the fruits over the sands [1]) or by water on which the fruits and the seeds can float. From an experiment carried out by Schimper [2]) it appeared that the seeds can remain floating for 10–60 days on a

[1]) As Schimper [Indo-Malayische Strandflora (1891), p. 157] already observed they are far too big to be carried by the wind through the air across a broad sea.

[2]) Schimper, Indo-Malayische Strandflora (1891), p. 165.

$3^1/_2$% solution of NaCl. Guppy[1]) found that only half of the seeds examined by him possessed floating power. Hillebrand[2]) ascribes to the plant viscous seeds and fruits, which may adhere to the feathers of birds. But already many years ago Guppy[3]) quite rightly observed that only the unripe fruits are sticky. None of the many hundreds of ripe fruits of this species which I saw during the last twenty years on the beaches of Java was in the least viscous, neither are the seeds sticky. Therefore I cannot in the least agree with the statement of Docters van Leeuwen[4]), that this plant may be spread also by animals, to which statement in the Dutch text the erroneous explanation is added „viscous seeds".

Two forms of this polymorphous species occur in Java, a pubescent arboreous one, growing in the eastern half of the island between 750 and 2750 m. above sea-level and a glabrous shrubby one spread from West- to East-Java on the beach. Of this last form a few specimens were found on young parts of the beach.

Malvaceae.

29. *Hibiscus tiliaceus L.* See p. 93, No. 11.
A few young specimens on the littoral plain.

Guttiferae.

30. *Calophyllum inophyllum L.* See p. 44, No. 6.
A few young specimens on the littoral plain.

Lecythidaceae.

31. *Barringtonia asiatica Kurz* (B. speciosa Forst.). See p. 43, No. 5.
Several very young specimens on the younger parts of the beach.

Combretaceae.

32. *Terminalia catappa L.* See p. 39, No. 2.
A few young specimens.

---

1) Guppy, Observations of a Naturalist in the Pacific between 1896 and 1899, II (1906), 339.
2) Hillebrand, Flora of the Hawaiian Islands (1888).
3) Guppy, Observations of a Naturalist in the Pacific between 1896 and 1899, II (1906), 340.
4) Docters van Leeuwen in Handelingen Eerste Natuurwetenschappelijk Congres (1920), p. 60; in Ann. Jard. Bot. Buitenzorg XXXI (1921), p. 129.

Melastomataceae.

33. *Melastoma malabathricum L.* (M. polyanthum Bl.).

One of the commonest Javanese shrubs, spread throughout the island from the plains up to ± 1650 m. above sea-level, except in regions where the east-monsoon is of very long duration. The plant grows by preference in sunny or slightly shadowed not too dry localities, in grassy jungles, thickets and young secondary forests, locally often in great numbers. The flowers [1] are adapted to pollination by insects, but I cannot adhere to the opinion of Fritz Mueller and Forbes that these animals always bring about *cross*-pollination. As a plant often bears many opened flowers at a time, geitonogamy can quite as well take place. Moreover it seems not yet to have been examined whether fructification *depends* on pollination by insects though such is probably the case. When ripe the oblong fruits burst transversely; the upper part claps back as a lid, denuding the much swollen juicy blackish red placentae which bear on their surface numerous minute seeds. These placentae are eaten by children and by animals which possibly distribute the seeds. Perhaps rain-water also takes a part in the distribution of these.

Several specimens in crevices at the base of the basaltic rocks.

Apocynaceae.

34. *Cerbera manghas L.* (C. odollam Gärtn.). See p. 46, No. 3.

A few very young seedlings on the younger parts of the beach.

Convolvulaceae.

35. *Ipomoea pes caprae Sw.* See p. 48, No. 7.

Numerous, in many places covering the beach with a loose network of stems. See Ernst, Neue Flora Vulkaninsel Krakatao (1907) tab. VII, fig. 4.

Verbenaceae.

36. *Premna integrifolia L.* (Pr. foetida Reinw.). See p. 137, No. 37.

A few scattered specimens on the littoral plain.

---

[1] Forbes, A Naturalist's Wanderings in the Malayan Archipelago, p. 228 seq.

37. *Clerodendron inerme Gärtn.*

Erect, sarmentose or subscandent shrub whose ripe fruits[1]) possess a thick corky exocarp, separating into 4 parts, each of which contains a hard one-seeded kernel. The plant grows on the beach and especially along river-mouths and on other water-sides near the sea; one often sees its fruits floating on the water. Ernst supposes[2]) that the fruits are also spread by birds, but this I think very improbable.

A few scattered specimens on the littoral plain.

Labiatae.

38. *Hyptis brevipes Poit.* See p. 138, No. 38.

Several specimens at the base of the basaltic rocks.

Rubiaceae.

39 *Morinda citrifolia L.* See p. 93, No. 16.

Several young specimens on the littoral plain.

Goodeniaceae.

40. *Scaevola frutescens Krause* (Sc. Koenigii Vahl). See p. 49, No. 9.

Several specimens, also on the somewhat older parts of the beach.

Compositae.

41. *Vernonia cinerea Less.* See p. 118, No. 58.

Several specimens on the older part of the littoral plain.

42. *Erigeron linifolius Willd.* (In Ernst's paper under the names of Blumea hieracifolia D.C. and Conyza angustifolia Ham.). See p. 51.

A few specimens at the base of the steep wall which secludes the littoral plain from the interior.

43. *Blumea lacera D. C.* See p. 51.

Several specimens on and near the base of the basaltic rocks.

44. *Erechthites hieracifolia Rafin.* See p. 52.

A few specimens in crevices at the base of the basaltic rocks.

---

[1]) Described by Schimper, Indo-Malayische Strandflora (1891), p, 173. The statement l. c. p. 75 that the fruit of this species is a berry, is incorrect.

[2]) Ernst, Neue Flora Vulkaninsel Krakatau (1907), p. 15.

45. *Pluchea indica Less.* (Conyza indica Miq.). See pp. 50, 108 and 119.

Here and there on the oldest parts of the littoral plain at the base of the rupture.

46. *Wedelia biflora D. C.* (In Ernst's paper named Wedelia glabrata Benth. and Wollastonia spec.). See p. 49 under Wollastonia.

Several specimens, especially on the younger parts of the beach.

B. FINDS ON THE BEACH ON THE SOUTH-EASTERN SIDE.

The character and composition of the littoral flora on the south-eastern side of Krakatao in 1906 are excellently described by Ernst [1]) to whose paper I refer. A few unimportant inaccuracies will be rectified herebeneath [2]).

The beach, consisting of a mixture of pumice, ashes and sand, is being washed away in many spots by the sea and wherever this is the case, slopes rather strongly downwards; in other localities it increases. Hence the boundary between beach and interior is often not clearly demarcated; a locality at present belonging to the interior may shortly afterwards, in consequence of land being rapidly washed away, border on the sea and is then liable to be confounded with a beach though, oecologically, it differs much from it and bears quite another, a halophobous vegetation. On the other hand, where a beach more rapidly increases than it is leached out by the rains, littoral halophytes may be found at a rather large distance from the sea. Yet these halophytes are usually in course of time replaced by other plants, able to live on the former beach as soon as this has lost its excess of salt by leaching caused by repeated rains or by water seeping through the sand in a lateral direction. *Consequently the life of many littoral plants is endangered as well by increase as by decrease of the beach.* Though the beach on the south-eastern side of Krakatao is of rather varied aspect in consequence of increase and decrease taking place, it does not possess the steep wall and the basaltic rocks of Zwarte Hoek. This explains that the last-named locality though only

---

1) Ernst Neue Flora Vulkaninsel Krakatau (1907), p. 28 seq.

2) Ernst [Neue Flora Vulkaninsel Krakatau (1907), 30] wrongly mentions *Caesalpinia crista L.* (C. bonducella Flem.) and *Pemphis acidula Forst.* as having been found in 1906 on the beach of the south-eastern side. For *Tournefortia argentea L. fil.* see the list beneath sub No. 30.

± $^{3}/_{4}$ km. in length and less densely covered with plants yielded in 1906 more species than the beach on the south-eastern side which was investigated over a length of at least $1^{1}/_{2}$ km. in a northern direction. Whilst at Zwarte Hoek 46 species were collected, we found only 36 on the beach of the south-eastern side. These were :

Cycadaceae.

1. *Cycas Rumphii Miq.* (in Ernst's paper named C. circinalis L.) See p. 130, No. 7.
   A female specimen of medium size (height 1,65 m.; circumference of the trunk 80 cm.) on the border of a *Casuarina*-grove somewhat behind the flood-mark. In spite of diligent searching we found no second specimen on the beach, but as this species is not restricted to the beach but may occur some kilometers remote from it, it is not quite impossible that other specimens were present here or there in the lower parts of the interior.

Pandanaceae.

2. *Pandanus tectorius Sol.* (P. littoralis Jungh.) See p. 43, No. 4.
   Several specimens just behind the flood-mark. See the fine picture in Ernst, Neue Flora Vulkaninsel Krakatao (1907), tab. V. fig. 9.

Gramineae.

3. *Ischaemum muticum L.* See pp. 89 and 90.
   In many places on the beach.

4. *Spinifex littoreus Merr.* (Sp. squarrosus L.). See pp. 89 and 90.
   In many places forming groups on the beach.

Cyperaceae.

5. *Cyperus pennatus Lmk.* (In Ernst's paper named Cyperus digitatus Roxb. and Mariscus umbellatus Vahl.).

6. *Fimbristylis spathaeea Roth.*
   See for these two species p. 47, No. 5 and 6; p. 90 and p. 91, No. 4 and 5. Both were found here and there on the beach, especially on the older parts.

7. *Remirea maritima Aubl.* See p. 47 and p. 132, No. 12.
   A few specimens on a young part of the beach.

Palmae.

8. Cocos *nucifera L.* See p. 39, No. 3.

A few specimens on a part of the beach which was being washed away (See Ernst, Neue Flora Vulkaninsel Krakatao (1907), p. 32, tab. V, fig. 8 [1])]. For other specimens, see list C, No. 11.—Here also many empty coco-nut shells were found in the drift-zone.

Casuarinaceae.

9. *Casuarina equisetifolia Forst.* See p. 91, No. 6.

Numerous on the south-eastern angle, forming there a rather large grove. According to Ernst the tallest specimens were 12—15 m. high, but I think they were rather higher (up to at least 20 m.). Farther to the north this species occurred in scattered specimens only.

Lauraceae.

10. *Cassytha filiformis L.* See p. 91, No. 7.

Numerous, forming its webs as well in the *Pes-caprae*-formation as over shrubs and low trees immediately behind the flood-mark. Ernst's statement [3]) that this species, together with other plants, formed „förmliche Dickichte" along the borders of the gullies of the interior is incorrect.

Hernandiaceae.

11. *Hernandia peltata Meissn.* See p. 47, sub No. 4.

A few scattered specimens.

Leguminose.

12. *Sophora tomentosa L.* See p. 147, No. 21.

A few scattered specimens.

13. *Indigofera zollingeriana Miq.* (In Ernst's paper named I. Teysmanni Miq).

---

[1]) Dr. S. Leefmans [2]) the well-known Buitenzorg entomologist, who was among the members of the congress who visited Krakatao in Oct. 1919, found on that occasion at Zwarte Hoek some specimens of Cocos *nucifera* which had been rather badly damaged by *Oryctes rhinoceros L.*, a beetle which is a serious pest to this palm in Java and Sumatra, killing many trees. According to Dr. Leefmans the leaves of the specimen of Cocos *nucifera* figured on the above mentioned table of Ernst show injuries distinctly pointing to *Oryctes*. Hence he thinks it probable that in 1906, during Ernst's visit, *Oryctes* was already present on the island.— If so, some of the trees found in that year may afterwards have fallen victims to the beetle.

[2]) S. Leefmans, Mededeelingen van het Instituut voor Plantenziekten (te Buitenzorg), No. 41, De Klappertor (*Oryctes rhinoceros L.*), 1920, p. 53.

[3]) Ernst, Neue Flora Vulkaninsel Krakatau (1907), p. 32.

Erect shrub or small tree, from West- to East-Java occurring here and there on sandy or rocky beaches, always in scattered specimens. It has also been found in the interior but it is possible that these specimens were cultivated ones or the offspring of these. The reproduction is effected exclusively by seeds. The pods do not burst when the seeds are ripe but fall off as a whole; they possess floating power.

Some scattered specimens.

14. Desmodium umbellatum D. C. See p. 148, No. 22.

Several specimens.

15. Pongamia pinnata Merr. (P. glabra Vent.).

Medium-sized tree, spread from West- to East-Java on sandy or rocky coasts, especially occurring in the *Barringtonia*-formation and there often frequent. The rather thick, coriaceous, indehiscent pods [1]) contain a single seed, not entirely filling up the fruit-cavity; they possess floating power and are often washed ashore.

Some scattered specimens.

16. Erythrina variegata L. var. *orientalis Merr.* (E. indica Lmk.) See p. 46, sub No. 1.

Here and there on the beach.

17. Canavalia rosea D. C. (In Ernst's paper named C. obtusifolia D. C.). See p. 89 and p. 92, No. 8.

In many places on the beach. By some mistake Ernst [2]) mentions this creeper among the twiners which formed a „fast lückenlose Laubwand" with young Casuarinas and shrubs. Incorrect is likewise Ernst's statement [3]) that this species together with other plants, formed „förmliche Dickichte" along the borders of the gullies of the interior.

18. Vigna marina Merr. (V. lutea Bth. and V. luteola Bth.). See pp. 89 and 93.

In many places creeping over the beach and twining in

1) Described by Schimper, Indo-Malayische Strandflora (1891), p. 164.

2) Ernst, Neue Flora Vulkaninsel Krakatau (1907), p. 30. Among these twiners *Caesalpinia crista L.* (in Ernst's paper C. bonducella) is also recorded and likewise wrongly so, for though a climber it is no twiner and as Ernst rightly observed on p. 39 this species was not found in 1906. It is not mentioned for Krakatao in the list of finds given by Ernst (p. 42), but nevertheless given for the interior on p. 55.

3) Ernst, Neue Flora Vulkaninsel Krakatau (1907) p. 32.

the shrubs immediately behind it. Farther into the interior no specimens were seen. Ernst's statement[1]) that this species together with other plants, formed „förmliche Dickichte" along the borders of the gullies of the interior is incorrect.

Meliaceae.

19. Xyocarpus granatum Koen. (Carapa obovata Bl.).

Small crooked tree with horizontal sinuous, laterally much compressed, respiratory roots, the upper part of which emerges from the muddy soil like the blade of a blunt knife. It is spread from West- to East-Java in the tidal forests especially in the less salt parts of these; it is locally often very common, though absent in many places. It frequently grows along the rivers and creeks intersecting the mangrove, and then usually on the convex (when seen from the middle of the river), i.e. the accrescent side of the curves. This tree and *Heritiera littoralis Dryand* are in the mangrove-forests of Java the only two species of which the trunks are often densely clothed with epiphytes whilst these are lacking on most other mangrove-trees. The reproduction of *Xylocarpus granatum* is effected by seeds which fall out when the large, subglobose fruits burst. They possess a very thick spongy testa, hence floating power [2]).

Only one specimen was seen.

Euphorbiaceae.

20. Euphorbia atoto Forst. See p. 90 and p. 93, No. 10.

In many places.

Sapindaceae.

21. Dodonaea viscosa Jacq. See p. 148, No. 28.

Several specimens.

Vitaceae.

22. Columella trifolia Merr. See p. 136, No. 28.

A few specimens, climbing high into Casuarinas near the beach.

---

1) Ernst, Neue Flora Vulkaninsel Krakatau (1907) p. 32.

2) Miquel (Flora Ind. Bat. I², 546), and on his authority also Guppy [Observations of a Naturalist in the Pacific between 1896 and 1899, II (1906), 564] suppose that the seeds of this genus germinate already in the fruit. With none of the very numerous specimens I have seen in Java this was the case. Neither has Schimper (Indo-Malayische Strandflora, p. 43) observed viviparity in this genus.

Malvaceae.

23. *Hibiscus tiliaceus L.* See p. 93, No. 11.
Here and there immediately behind the floodmark.

Guttiferae.

24. *Calophyllum inophyllum L.* See p. 44, No. 6.
Several scattered specimens. See also list C, No. 19a.

Lecythidaceae.

25. *Barringtonia asiatica Kurz.* See p. 43, No. 5.
Several scattered specimens. See also list C, 19 b.

Combretaceae.

26. *Terminalia catappa L.* See p. 39, No. 2.
Several scattered specimens.

Convolvulaceae.

27. *Ipomoea denticulata Chois.*

Twining or creeping and then copiously rooting, rather slender herb, not reaching a great height, occurring in Java especially in the western half, exclusively on sandy beaches and the localities close to these. The reproduction is effected by the seeds which possess floating power.

Several specimens especially near the south-eastern point of the island.

28. *Ipomoea longiflora R. Br.* (I. grandiflora Lmk.; in Ernst's paper under the name of Calonyction asperum Chois.).

Twining herb of the *Barringtonia-* and *Pes-caprae*-formations especially in West-Java of common occurrence. The reproduction is effected by the fruits, which are enclosed within the calyx which, after the anthesis, becomes much enlarged and secerns slimy water on its inner side [1]). Each capsule contains 1–4 seeds; the bifid cotyledons enclose between their curved halves a rather large air-cavity. Hence the seeds possess floating power; not rarely they are found washed ashore. Guppy found them in the drift-zone of the Keeling islands and observed that the seeds of fruits which had been

[1]) This species was not yet mentioned by Nils Svedelius in his paper: Ueber das postflorale Wachstum der Kelchblätter einiger Convolvulaceen (Flora XCVI (1906 p. 231 seq.).

floating for a year on sea-water, were still fit to germinate [1]) Docters van Leeuwen guesses [2]) that the seeds are also spread by birds but gives no grounds for that opinion. I cannot believe that the seeds of any species of *Ipomoea*, if eaten at all by birds, pass their intestine tract without being destroyed.

Several scattered specimens.

29. *Ipomoea pes caprae* Sw. See p. 48, No. 7.
Numerous.

Borraginaceae.

30. *Tournefortia argentea L. fil.* See p. 54, No. 8.

Ernst [3]) states that this shrub was found in 1906 on the south-eastern side of Krakatao. I did not see it there neither in 1906 nor in 1908, but found it in 1906 on Verlaten Eiland. But as both Treub in 1886 and Docters van Leeuwen in 1919 found the plant on Krakatao I admit that it might have grown there in 1906 and have been overlooked by me. I must however decidedly contradict Ernst's statement [3]) that it formed together with other plants „förmliche Dickichte" along the borders of the gullies in the interior.

Verbenaceae.

31. *Premna integrifolia L.* (Pr. foetida Reinw.). See p. 137, No. 37.
Scattered specimens.

32. *Clerodendron inerme Gärtn.* See p. 151, No. 37.
A few scattered specimens.

Rubiaceae.

33. *Guettarda speciosa L.* See p. 138, No. 41.
A few scattered specimens.

34. *Morinda citrifolia L.* See p. 93, No. 16.
A few scattered specimens.

---

1) Guppy, Observations of a Naturalist in the Pacific between 1896 and 1899, II 531.
2) Handelingen Eerste Nederlandsch Indisch-Natuurwetenschappelijk Congres (1920), p. 56, – Ann. Jard. Bot. Buitenzorg XXX (1921), p. 127.
3) Neue Flora Vulkaninsel Krakatau (1907), p. 32.

Goodeniaceae.

35. *Scaevola frutescens Krause* (Scaevola Koenigii Vahl). See p. 49, No. 9.
In several places. Ernst's statement [1]) that this shrub formed together with other plants „förmliche Dickichte" along the border of the gullies in the interior is incorrect.

Compositae.

36. *Wedelia biflora D. C.* (In Ernst's paper named Wedelia glabrata B. et H., – Wedelia scabriuscula Boerl. and Wollastonia spec.). See p. 49, sub No. 1.
In many places.

C. FINDS IN THE INTERIOR.

The locality which was buried by the eruption of 1883 under a mixture of ashes and pumice rises near the sea very gently upwards towards the cone of the mountain, which lies north-west of our landing-place. Quite near the coast the plain is interrupted only by a number of broad and shallow gullies, which come out on the sea. These gullies have been eroded by rain-water flowing down; they contain, like the wadies of North-Africa, water only during and shortly after heavy showers. At a short distance from the coast they increase in number and depth; their steep walls consisting of loose materials can be climbed only with much difficulty and frequently not at all. The investigated part of the interior was in 1906 mainly clothed with a very dense, 2–5 m. high *Saccharum*-jungle [2]) which left only here and there some space for other plants and could be traversed only by the aid of a chopping-knife and at much pains. The endless difficulties presented by the unevenness of the locality, by the incredibly dense vegetation, by the tropical heat, by the lack of shadow and water, by the incessant attacks of innumerable vicious ants and mosquitoes, were cause that our progress was but very slow. When after hours of toilsome struggle[3]) we had at last to reluctantly admit that the short time still at our disposal did not allow us to continue our way, we were at best $1^1/_4$ km., as the crow flies, from the coast and still far from the lower wooded ravines. We had not reached a greater height than at best 40 or 50 m.,

---

1) Ernst, Neue Flora Vulkaninsel Krakatau (1907), p. 32.
2) See also p. 100 seq.
3) See Ernst, Neue Flora Vulkaninsel Krakatau (1907), pp. 32–34, already partly cited herebefore, p. 102.

perhaps rather less. Almost continously immured between two high walls of *Saccharum,* that hid from view everything not quite near, we could investigate only the flora of an exceedingly small part of the lower slopes, viz. the margins of the path cut by us and a few small open spots we happened to come across; we were not able to reach the higher ravines which were so much more important for the exploration. It is therefore quite out of the question that this short trip should have given a somewhat complete survey of the vegetation of the island in 1906: tens, perhaps hundreds of species may have remained unobserved. If one from the fact that in subsequent years species were collected on Krakatao not seen there in 1906, should conclude that all these species had established themselves there after that year, one would give proof of serious lack of insight into problems as that in question. To tell the sad truth, the botanical investigation of the interior in 1906 was a failure. For this failure the investigators, who did all they could, cannot be blamed, but nevertheless a failure it was. We observed the following plants:

Schizophyceae.

Bluish green algae covering pieces of pumice. Names unknown.

Bryaceae.

1. *Bryum coronatum Schwaegr.*

Calciphilous [1]) caespitose moss, living on (mostly calciferous) earth, brick walls and calcareous rocks, spread from West- to East-Java from the plains to an altitude of more than 1000 m.

On the lower south-eastern slopes forming tufts along the borders of the gullies.

Bartramiaceae.

2. *Philonotis secunda Van den Bosch et Lacoste.*

Moss, forming dense sods on rocks and earth, spread in West-Java from the plains to an altitude of ± 1500 m.

---

[1]) For the CaO-percentage of part of the eruptive products see p. 59, sub No. 9, *Pteris aquilina L.* This same moss was found i.a. on the calcareous rocks of Kuripan near Buitenzorg. [Cf. Moritzi, Systematisches Verzeichniss der von H. Zollinger in den Jahren 1842–1844 auf Java gesammelten Pflanzen (1845–46), p. 133, under the name of *Br. doliolum Duby*].

On the lower south-eastern slopes forming tufts along the borders of the gullies.

Polypodiaceae.

3. *Nephrolepis exaltata Schott.* See p. 56, No. 6.
   Here and there on open spots in the grass-jungle.

4. *Dryopteris unita O. K.* (Nephrodium cucullatum Baker). See p. 56, sub No. 7.
   In many places on steep walls and in open, stony spots in the grass-jungle.

5. *Ceropteris calomelanos Und.* See p. 54, No. 1.
   In many places on steep walls of ravines and gullies.

Gramineae.

6. *Saccharum spontaneum L.* See p. 100.
   From close to the floodmark to far into the interior growing gregariously, forming an almost impenetrable jungle, by far the most common plant in the investigated parts behind the coast.

7. *Pogonatherum paniceum Hack.* (Pogonatherum crinitum Trin.). See p. 105.
   In many localities on steep walls.

8. *Pennisetum macrostachyum Brongn.* (Gymnothrix elegans Büse; Pennisetum elegans Ernst). See p. 48, No. 8 and p. 99.
   Here and there scattered in the *Saccharum*-jungle and along the borders of the gullies, nowhere gregarious.

8a. *Neyraudia madagascariensis Hook. fil.* var. *Zollingeri Hook. fil.* See p 103.
   On p. 32 of his paper [1]) wrongly mentioned by Ernst as found 1906 on Krakatao, but in his list of finds on p. 44 rightly omitted.

Cyperaceae.

9. *Cyperus pennatus Lmk.* (In Ernst's paper named C. digitatus Roxb. and Mariscus umbellatus Vahl.).

10. *Fimbristylis spathacea Roth.*
   See for these two species p. 47, No. 5 and 6; p. 90 and p. 91, No. 4 and 5.

---

1) Ernst, Neue Flora Vulkaninsel Krakatau (1907), under the name of *Phragmites Roxburghii Steud.*

On the beach and somewhat behind it in scattered tussocks; more in the interior not seen.

Palmae.

11. *Cocos nucifera L.* See p. 39, No. 3.

A grove of several specimens rather far [1]) from the beach and a few meters above the floodmark. Several of the trees bore ripe fruits; on the ground beneath many fallen nuts were lying, some of which had germinated. We also found some nuts that had been cut open, which proves that these trees had already been visited by man [2]). The older trees of the group seemed, by their size, to have reached an age of about ten years. Nearer to the beach some scattered specimens of this species were found.

Orchidaceae [3]).

12. *Arundina speciosa Bl.* See p. 107.

Here and there in open stony places in the *Saccharum*-jungle.

13. *Spathoglottis plicata Bl.* See p. 107.

Here and there on ridges and steep walls.

14. *Cymbidium Finlaysonianum Lindl.*

Orchid usually living on trees, very rarely on the ground, found here and there in West- and Central-Java beneath an altitude of 500 m. in sunny or moderately shadowed localities, on trees planted along road-sides, in teak-forests, in thin

---

1) According to Ernst [Neue Flora Vulkaninsel Krakatau (1907), p. 70] this group was ± 400 m. distant from the sea, but according to the maps in Jaarverslag van den Topographischen Dienst in Ned.-Indië (Year-book of the Topographical Survey Service in the Dutch Indies), 1908, plate XXIII and in Handelingen van het Eerste Nederlandsch Indisch Natuurwetenschappelijk Congres (Transactions of the First Dutch Indian Congress of Natural Science 1920) about 75 m. It seems to me that the last given distance is too short, but that given by Ernst too long: between 1906 and 1908 the land cannot have been washed away over a breadth of more than 300 m. I for me estimate that the coco-nut grove was about 150—200 m. distant from the sea.

2) This fact was already mentioned by me in Jaarboek van het Departement van Landbouw in Nederlandsch Indië (Year-book of the Department of Agriculture in the Dutch Indies] 1906, p. 4. Valeton does not mention to have found coco-palms in 1905. As it is not known that any European has visited the island between the trips of 1905 and 1906 it is probable that native fishermen had consumed the contents of the nuts, as they are wont to do.

3) According to Ernst [Neue Flora Vulkaninsel Krakatau (1907), p. 33] besides the species recorded above also a species of *Phajus* was found in the interior of Krakatao in 1906. But in his list of finds on pp. 43, 44 he does not mention this genus for the island. I myself did not see it there in 1906 but found it in 1908 in ravines at an altitude of 200 m. and more.

rain-forests and in the mangrove, generally in scattered specimens. In Sept. 1896 a sterile specimen of this species was found by Boerlage on Lang Eiland. This specimen (No. 71 of his finds) was wrongly named by him *Vanda*[1]). In later years this error was corrected by Dr. J. J. Smith.

From investigations made by Forbes[2]) it may perhaps be concluded that *Cymbidium Finlaysonianum* is not autogamous. This may be one of the causes that in 1906 this orchid was much less frequent in the investigated parts of Krakatao than *Arundina* and *Spathoglottis*, which are frequently autogamous (See p. 108).

One specimen of this orchid was found on a steep ravine-wall.

Casuarinaceae.

14a. Casuarina equisetifolia Forst. See p. 91, No. 6.

Though some scattered specimens of this tree were found quite near the *Casuarina*-grove mentioned herebefore (p. 154, No. 9) somewhat farther from the coast, there is no reason to consider this tree as an inland species.

Moraceae.

15. Ficus fistulosa Reinw. See p. 133, No. 16.

A few specimens, some of them bearing unripe figs, in grass-jungles near the beach.

16. Ficus fulva Reinw. See p. 133, No. 17.

A few specimens, some of them bearing unripe figs, in grass-jungles near the beach.

---

1) Whether the orchid found in March 1897 by Penzig on Lang Eiland which was named by Boerlage *Vanda sulingi Bl.* and recorded under this name in Penzig's list [3]), was also this *Cymbidium* I cannot decide, the specimen not being present in the Buitenzorg Herbarium. Anyhow, it was certainly **not** *Vanda sulingi* (at present named *Arachnis sulingi Rchb. fil.*) for this epiphyte does not grow on the beach but is restricted to shadowy forests and in 1897 the explorers visited on Lang Eiland only the beach [4]) where no shadowy forests were seen but only a small grove of Casuarinas, 5–6 m. high. Because in Penzig's list all data on habitat, frequency, a. s. o. are lacking, only an examination of the specimen in question can bring the decision, which, at present, is of little or no importance for us.

2) Forbes, A Naturalist's Wanderings in the Eastern Archipelago (1885) p. 83.

3) Ann. Jard. Bot. Buitenzorg XVIII (1902), p. 104.

4) Ann. Jard. Bot Buitenzorg XVIII (1902) p. 107. All subsequent authors have recorded Penzig's plant either as *Vanda sulingi* or *Arachnis sulingi*, proving by doing so, that a vegetation cannot be well described by authors for whom the plants they write upon are no living things but Latin names.

Urticaceae.

17. Pipturus incanus Wedd. See p. 134, No. 19.
Some scattered specimens on open spots in the *Saccharum*-jungle.

Leguminosae.

18. Cassia siamea Lamk.
Tree, a native of British India and Cochinchine, in Java very often cultivated as a shade-tree along road-sides or on premises and not rarely running wild. The reproduction is effected exclusively by seeds.
One very young specimen quite near the flood-mark.

Sterculiaceae.

19. Melochia umbellata Stapf (Melochia arborea Blanco; M. indica A. Gray). See p. 117, No. 51.
Here and there in open places in the *Saccharum*-jungle.

Guttiferae.

19a. Calophyllum inophyllum L. See p. 44, No. 6.
A few specimens somewhat behind the beach. Probably these specimens were growing once immediately behind the flood-mark and got removed from it by an increase of the beach. There is no reason to consider this tree as an island species.

Lecythidaceae.

19b. Barringtonia asiatica Kurz (B. speciosa Forst.). See p. 43, No. 5.
A few specimens somewhat behind the beach. Probably these specimens were growing once immediately behind the flood-mark and got removed from it by an increase of the beach. There is no reason to consider this tree as an inland species.

Borraginaceae.

19c. Tournefortia argentea L. fil.
See p. 158, No. 30.

Cucurbitaceae.

20. Trichosanthes bracteata Voigt (In Ernst's paper under the name of Tr. tricuspidata Lour.). See p. 139, No. 43.

A few specimens somewhat behind the beach, climbing in *Ficus*-trees (No. 15 and 16 herebefore).

Compositae.

21. Blumea balsamifera D. C. See p. 51.

A few specimens in somewhat open spots in the grass-jungle. In 1906 I did not observe any other Compositae in the interior on the S. E. side. Ernst [1]) mentions to have seen „einige gelbblühende Compositae", but all these may have belonged to this species.

D. PLANTS MENTIONED BY ERNST, THE EXACT HABITAT OF WHICH IS UNKNOWN TO ME.

Bacteriaceae.

1. Bacterium Krakataui De Kruyff.

Aerobic nitrogen-fixing bacterium isolated by De Kruyff [2]) from soil-samples collected by Ernst in the *Casuarina*-grove (on the south-eastern angle) and in the *Pes-caprae*-formation (where ?).

2. *Bacillus mycoides Fluegge.*

3. *Bacillus Moiré* [3]) (B. megatherium De Bary ?).

These two species were isolated from soil-samples [4]) collected either in the *Barringtonia*-formation, the *Casuarina*-grove, the littoral zone or the interior. Very numerous.

4. *Bacillus radicicola Beyerinck.* (Bacterium radicicola Beyerinck).

Diverse Leguminosae of the beach (*Canavalia, Erythrina, Vigna*) bore on their roots nitrogen-tubercles.

5. *Pseudomonas fluorescens Migula* (In Ernst's paper named Bacillus fluorescens liquefaciens Flügge).

This species was isolated from soil-samples collected either in the *Barringtonia*-formation, the *Casuarina*-grove, the littoral zone or the interior.

---

1) Ernst, Neue Flora Vulkaninsel Krakatau (1907), p. 33.

2) E. De Kruyff, Sur une bactérie aérobe fixant l'azote libre de l'atmosphère, in Bull. Dép. Agric. Ind. Néerl. (1906), p. 9.

3) I have not been able to find back this singular name in the systematic litterature at my disposal.

4) „In allen 4 Proben fanden sich ferner Erreger der Cellulose-, Pektinstoff- und Stärkegärung, sowie Urobakterien. Auf der Papierplatte" (Methode Van Iterson) „kamen zahlreiche Schimmelpilze zur Entwicklung" (Ernst, Neue Flora Vulkaninsel Krakatau (1907), p. 53).

Polyporaceae.

6. *Polystictus hydnoides Berk.*
Found on a tree-trunk washed ashore.

7. *Pycnosporus sanguineus Murrill* (Polystictus sanguineus Fries).
A fungus of decaying wood, very common in Java [1]. Found on a tree-trunk washed ashore.

Polypodiaceae.

8. *Pteris vittata L.* (In Ernst's paper named Pt. longifolia L.) See p. 56, No. 5.
In 1906 I saw this species on Verlaten Eiland but not on Krakatao.

9. *Acrostichum aureum L.* See p. 55, No. 4.
In 1906 not seen by me, neither on Krakatao nor on Verlaten Eiland.

Leguminosae.

10. *Caesalpinia crista L.* (In Ernst's paper named C. Bonducella Flem.) See p. 155, footnote 2.

Anacardiaceae.

11. *Spondias mangifera Willd.*
Tree, spread from West- to East-Java from the plains up to an altitude of ± 600 m., especially growing in regions where the east-monsoon is rather strong, on periodically dry soils, often in teak-forests and along the inner borders of the mangrove; it does not need a fertile soil. The reproduction is effected by the rather large juicy drupes. The woody kernel of these is provided with 5 much prominent longitudinal ridges, the interstices between which are filled up with a fibrous tissue; hence the kernels possess floating power. The fruits are eaten by animals, i.a. bats.

In 1906 not found by me on Krakatao but I collected one young specimen on the beach of Verlaten Eiland for which island the species is not recorded by Ernst.

---

[1] A description and a fine coloured plate by Dr. C. van Overeem († 1927) were published in De Tropische Natuur XIII (1924), p. 83.

Asclepiadaceae.

12. Cynanchum laeve Schum. (C. Blumei B. et H.).

Twiner, found in West-Java in several localities between 20 and 1000 m. above sea-level on forest borders and in thickets. The reproduction is effected by the comose seeds which over land may be easily spread by the wind over some distance. The ripe seeds very easily let loose from the tuft of hairs.

A fruiting specimen collected in 1906 by Ernst on or near the beach of Krakatao [1]) is preserved in the Buitenzorg Herbarium. As flowers are lacking the determination is not quite certain.

At the end of his description of the excursion of 1906 Ernst gives a list of all plants found till that year on the 3 islands of the Krakatao-group. Recording a few of these under different names twice or even three times he comes to a total of 137 species. Among these are 108 Phanerogams and Vascular Cryptogams of which 67 were found exclusively or mainly on the beach, the others in the interior. After making the necessary corrections and adding the plants found in 1905 by Valeton I come for Krakatao itself to 119 species, of which 90 are Phanerogams or Vascular Cryptogams. Among these are 39 or 40 littoral species and halophytes, the ± 50 others are plants of the interior [2]).

It is clear that in drawing conclusions from a comparison of the vegetation of Krakatao in 1905/1906 with that of 1897, all plants should left out of account which were found on the south-eastern side which was never visited by a botanist before 1905 and likewise all plants of which the habitat is unknown. On the other hand it should not be forgotten that in 1897 the slopes above Zwarte Hoek were investigated up to a not inconsiderable height whilst in 1905 Valeton went only some short way up; in 1906 these slopes, their very base excepted, were not explored at all Doubtless this is one of the causes that several plants, collected at former excursions, were not found back in 1906. For instance, in 1897 the number of ferns observed was 13, in 1906 but 5, among which not a single new one. From this fact may only be concluded that the exploration in 1906 was not sufficiently intensive. There is only one part of the

---

[1]) Ernst, Neue Flora Vulkaninsel Krakatau (1907), p. 47.

[2]) One should not forget that on unstable coasts like that of Krakatao it is often difficult to clearly distinguish between beach and interior (taken in the oecological sense of the words).

island for which a comparison of the vegetation in 1897 with that in 1906 is, to some extent, possible, viz. the littoral plain at Zwarte Hoek, unhappily one of the least important parts of the island, as it is certain that this plain did not exist before the eruption. And not even this comparison can be safely made, as Penzig did not mention the habitat of most of his finds. The older parts of the beach may during his excursion already have borne several halophobous non-littoral species. But only for one species, *Saccharum spontaneum L.* this is out of doubt. Of the rest we know nothing.

Restricting ourselves to littoral species and halophytes we find that Penzig found at Zwarte Hoek 18 of these (See the list on pp. 90 seq.). In 1906 all these plants were found back by Ernst; moreover the following 8 littoral or halophilous species were collected: *Pithecolobium umbellatum Bth.* (20) [1], – *Sophora tomentosa L.* (21), – *Desmodium umbellatum D.C.* (22), – *Derris heterophylla Backer* (23), – *Erythrina variegata L.* var. *orientalis Merr.* (24), – *Dodonaea viscosa Jacq.* (28), – *Calophyllum inophyllum L. (30)*, – and *Clerodendron inerme Gärtn. (37)*. Though fully admitting that young specimens of some of these 8 species may have been present in 1897 at Zwarte Hoek [2] but were overlooked, I can hardly believe that such has been the case with all of them, the beach having been investigated by 4 botanists and a very able native collector. Therefore I think we may take for granted that the beach at Zwarte Hoek was in 1906 richer in species than in 1897. But from this may by no means be concluded that it has been proven that the littoral flora of Krakatao, *as a whole*, had grown richer. From a comparison of the finds at Zwarte Hoek with those on the south-eastern beach it appears that in 1906 in the latter locality 9 or 10 littoral plants were collected that had not been found in the former, viz. *Cycas rumphii Miq.* (*1*) [3], – *Remirea maritima Aubl.* (7), – *Hernandia peltata Meissn.* (*11*), – *Indigofera zollingeriana Miq.* (*13*), – *Pongamia pinnata Merr.* (*15*), – *Xylocarpus granatum Koen.* (*19*), – *Ipomoea denticulata Chois. (27)*, – *Ipomoea longiflora R. Br.* (28) and *Guettarda speciosa L. (34)*. To this list should perhaps be added *Tournefortia argentea L. fil.* (*30*), which I myself did not see but

[1]) The numbers between brackets refer to list A herebefore.

[2]) Young specimens of *Calophyllum inophyllum L.* and of a species of *Erythrina* were already found by Treub in 1886. Evidently these seedlings did not grow up into adult trees, else Penzig could not possibly have overlooked them. I refer once more to what I said on page 44 about the dangers to which young seedlings on the beach are exposed.

[3]) The numbers between brackets refer to list B herebefore.

which may have been rightly recorded by Ernst for this side of the island. On the other hand 3 littoral species not observed on the south-eastern side were collected at Zwarte Hoek, viz. *Pithecolobium umbellatum Bth.* (*20*) [1], – *Derris heterophylla Backer* (*23*) and *Cerbera manghas L.* (*34*). This proves how dangerous it is to conclude from the investigation of *a part* of the beach, that a littoral species is absent *from the entire island,* and I cannot at all agree with Ernst who apparently considers all species observed for the first time in 1906 on Krakatao as „in den letzten 10 Jahren eingewanderte „Arten" [2]. All of these species may have occurred on Krakatao already long before 1897.

The typical littoral species and halophytes were most probably introduced after the eruption, but the same cannot be said of the flora of the interior, the total destruction of which in 1883 is not in the least proven. The vegetation of the lower parts of the slopes of Krakatao, like that of all localities clothed with an undisturbed young vegetation in moist regions of the tropics, has certainly in the course of years much changed. But we do not in the least know whether the „new" plants found in 1906 on these slopes were new for the island itself or had come over from other parts of it. All that has been said about the flora of the interior between 1886 and 1906 having grown richer, and on the mode of introduction of the so-called new inland species are *mere guesses* without scientific value.

I shall shortly discuss the species found on Krakatao for the first time in 1905 or 1906. Soil-bacteria (No. 1-3 and 5 of list D) proved to be as copious as in Buitenzorg. It is fully unknown when they have first appeared in the investigated localities. It has been often *alleged* but was never *proven* that the island was quite sterilized by the eruption of 1883. I think it quite possible, even very probable, that in the higher ravines of the island which had been covered only with a thin and not very hot layer of ashes and grit, a number of bacteria have survived the eruption and from there have gradually spread over the thickly and permanently covered lower parts. Of course introduction from without the island can also have taken place. Attention should be paid to the new aerobic free-living nitrogen-fixing *Bacterium Krakataui* which was found in the *Casuarina*-grove and in the *Pes-caprae*-formation and to *Bacillus radicicola,* the common bacterium of the nitrogen-tubercles on the roots of *Legumi-*

---

[1]) The numbers between brackets refer to list A herebefore.

[2]) Ernst, Neue Flora Vulkaninsel Krakatau (1907), 46.

nosea. I fully acknowledge the importance of these bacteria for the *littoral* vegetation of the island but Ernst greatly exaggerates where he says [1]: „Die Leguminosen sind für die Neu-Besiedelung „der Insel von ganz besonderer Bedeutung geworden. Sie sind jetzt „mit 16 Arten (Krakatao: 12) die 14 (Krakatao: 12) Gattungen ange- „hören, im besonderen in der *Pes-Caprae-* und *Barringtonia*-formation „des Strandes, durch *Vigna, Canavalia, Caesalpinia* und *Mucuna* auch „in der Binnenlandflora vertreten und übertreffen hier wie dort fast „alle anderen baum- und strauchartigen Blütenpflanzen an Individuen- „zahl". *This entirely misrepresents the situation in 1906;* on the contrary the *Leguminosae* were restricted to the beach and the locality immediately behind it, partly a former beach; they were not found farther remote from the coast than a few meters. *Caesalpinia* and *Mucuna,* cited by Ernst as *Leguminosae* of the interior, were in 1906 *not at all* found on Krakatao but only on Verlaten Eiland [2]) and the exploration of this last island in 1906 (of but *one hour's* duration; area of the island about $11^1/_2$ km^2.) was carried out only on some parts of the beach and the locality immediately behind it. Two years afterwards the same experience was again made on Krakatao: *not a single Leguminosa* was found in the interior proper, and as late as 1919 all *Leguminosae* found on Krakatao since the eruption, had been collected on or quite near the beach, except perhaps *Albizzia chinensis Merr.* (habitat and frequency *unknown,* correctness of determination doubtful), which was found in 1897 by Penzig but was never afterwards found back and *Mucuna acuminata D.C.* (habitat and frequency unknown) found in 1905 by Valeton. In 1919 *Aeschynomene indica L.,* a very common hygrophyte in the Javanese rice-fields, was found on Verlaten Eiland, but this plant also was growing quite near the sea. As late as 1919 *Leguminosae took practically no part in the vegetation of the interior.* Nothing being known about the flora of the interior of Krakatao before 1883, we cannot make out whether the present lack of *Leguminosae* has to be ascribed to lack or scarcity of these plants in the former vegetation. But the conclusion seems to be warranted that, except by the sea, introduction of *Leguminosae* does not or hardly ever take place, unless one must suppose that seeds are imported but do not grow up into plants because the pumice- and ash-fields of Krakatao are unfit for their development. But as far as our

[1]) Ernst, Neue Flora Vulkaninsel Krakatau (1907), p. 55.

[2]) This also appears from the list given by Ernst himself (Neue Flora Vulkaninsel Krakatau (1907), pp. 42, 43).

present knowledge goes this last supposition seems to be baseless. Perhaps the humidity of the climate is also of some influence. Most of the Indian *Leguminosae* are inhabitants of dry regions. Nevertheless even in regions where the east-monsoon is so feeble as in Buitenzorg several *Leguminosae* occur in a wild state (i.a. *Alysicarpus nummularifolius* D.C., — *Cassia Leschenaultiana* D.C. and *Tora* D.C. — *Crotalaria incana* D.C. and *striata* D.C., — *Desmodium heterophyllum* D.C., and *triflorum* D.C., — *Mimosa pudica* L., — *Phaseolus sublobatus* Roxb. and several others), some of these even in great numbers. As late as 1919 none of these species had been found on Krakatao.

The short time that in 1906 could be given to the exploration was insufficient for a close investigation of the vegetation of algae. Bluish green algae were seen in the interior on the south-eastern side but we do not know to which species they belonged. We know nothing of the vegetation of algae on the pumice- and ash-fields above Zwarte Hoek; probably it had been locally superseded by higher plants and was present only on naked walls and such-like localities. The two new *Polyporaceae* were growing on a tree-trunk washed ashore but this is of course no proof that they were *introduced* in this manner. Their spores may quite well have come after the tree. Many other fungi may in 1906 have been present in the humid higher ravines; we do not know whether they were introduced after the eruption and if so, in which manner; their spores may very well have been carried over by the wind, but there exists *no proof* for this, it is only *a plausible guess.* Seventeen of the species [1]) found for the first time in 1905 or 1906 are plants which, in the Dutch Indies, are generally found on the beach and the localities immediately behind it. All these species have a wide distribution along the coasts and may have grown on Krakatao long before 1906. Without exception they possess seeds or fruits fit to be spread by sea-currents, hence it is very probable that they were imported by the sea. The nutlets of *Hyptis brevipes* Poit. (in Java a very common *Labiata* of watersides and rice-fields) which in 1905 had been found on Krakatao by V a l e t o n (habitat not mentioned) and was collected in 1906 at the very base of the basaltic rocks, possess floating power. These nutlets or fruit-bearing

[1]) *Clerodendron inerme* Gärtn., — *Columella trifolia* Merr., — *Cycas rumphii* Miq., — *Derris heterophylla* Backer, — *Desmodium umbellatum* D.C., — *Dodonaea viscosa* Jacq., — *Guettarda speciosa* L., — *Indigofera zollingeriana* Miq., — *Ipomoea denticulata* Chois., — *Ipomoea longiflora* R. Br., — *Pithecolobium umbellatum* Bth., — *Pongamia pinnata* Merr., — *Premna integrifolia* L., — *Remirea maritima* Aubl., — *Sophora tomentosa* L., — *Thespesia populnea* Soland., — *Xylocarpus granatum* Koen.

plants or fragments of these, after having been carried by rivers to the sea, may have been transported by sea-currents to Krakatao, though also other modes of import are quite possible, f.i. with rice-straw used for the packing of bottles. This part of Krakatao has been frequently visited by Europeans and doubtless these have carried drinks with them, at least water, but probably also something better. *Mucuna acuminata Grah.* may also have been introduced by the sea.

It is very doubtful whether among the older specimens of Cocos *nucifera L.* there were really wild ones. Most of them were remote from the sea but the beach may have increased after the nuts arrived at their present places. Conversely the specimen found below the flood-mark on a decreasing part of the beach on the south-eastern side must have grown previously more in the interior. Anyhow, it is certain that not a single undoubtedly wild specimen was found except young seedlings growing under fruit-bearing trees that stood themselves above flood-level and probably had been planted (See p. 39 seq.). In the list of plants found in 1905 by V a l e t o n on Krakatao Cocos *nucifera* does not occur. — *Cymbidium Finlaysonianum Lindl.* was found in a single specimen but this orchid was collected as early as 1896 by B o e r l a g e on Lang Eiland. To my knowledge it has not been found back on Krakatao since 1906 but this is of course no proof that it has disappeared. Looking for rare plants on Krakatao is like looking for a needle in a hay-stack. *Cassia siamea Lamk.*, which was found in a single quite young specimen close to the sea, is a tree that in the lower regions of Java is very frequently cultivated along road-sides and locally has more or less run wild. One can only guess how its seeds may have reached Krakatao, perhaps a branch with ripe pods was washed ashore. Neither do we know how the comose seeds of *Cynanchum laeve Schum.* and the winged seeds of *Uncaria* and *Nauclea* have been carried over to the island; it is a very alluring hypothesis to ascribe the introduction of these plants to direct action of the wind but in the first place it is not certain that they were introduced at all and are not survivors of the eruption, in the second place other modes of introduction (f.i. by the sea) are as well possible. *Barringtonia insignis Miq.*, — *Ficus fistulosa Reinw.*, *fulva Reinw.* and *hirta Vahl*, — *Leea aequata L.*, — *Macaranga tanarius M.A.* — *Pipturus incanus Wedd.* — *Trema orientale Bl.* and *Trichosanthes bracteata Voigt*, all of which posses fleshy or juicy fruits and are in Java common plant of forests, forest-borders, jungles or thickets are either survivors of the old vegetation or were introduced after the eruption. If the last be the case, one is inclined to suppose that such has been done by frugivorous animals. But for these plants also we cannot do more than make guesses as regards the three cardinal points of the problem, the *fact*, the *time* and the *mode* of import, because there exists no unimpeachable evidence for direct *introduction* of seeds or fruits by birds or other animals (which should not be confounded with *further dispersion* over the island). Here, as everywhere in the Krakatao-problem, very many guesses have been made but exceedingly little has been proven. It is unknown *which* plants of the interior were imported before 1906, it is unknown *when* they were imported, it is unknown in *which way* they were imported. None of the Krakatao excursions has spread any other light on these points than the often misleading twilight of guesses.

In Chapter IV of his paper Ernst discusses the conditions of life on Krakatao, especially the soil-conditions. On very good grounds he argues that the soil of Krakatao was by no means so unfertile as Treub meant. One very important source of plant-food is not mentioned by Ernst, viz. the former vegetation of such higher parts of the mountain as were soon laid bare or almost so by erosion. This subject was already discussed on p. 73.

In Chapter V of his paper Ernst excellently sets forth which part water, wind and birds *may* have taken in carrying seeds, fruits or spores to Krakatao. Below I give the essential parts of his argument in an abridged form.

The seeds of littoral plants are almost exclusively spread by sea-currents: one finds them washed ashore everywhere. Experiments made with many of these seeds have shown that they can remain floating for a shorter or longer time (1 – 120 days) on sea-water or on a solution of NaCl without losing their germinative power. Even such seeds as can float only one or a few days may quite well have been transported by the sea to Krakatao which is less than 45 km. distant from the surrounding islands, because in the Sunda Straits, at least in the east-monsoon, sea-currents run, which continue in the same direction for 18 hours with a minimal velocity of 3 km. an hour. Seeds without floating power may be carried over in hollows or crevices of tree-trunks or blocks of pumice. Rivers swollen by heavy showers carry not seldom uprooted trees to the sea; part of the crown or of the root-system of these trees often emerges from the water; not salt-resistent fruits or epiphytes can make the voyage in this manner. The emerging part of the root-system often holds some soil in which living plants or seeds may be present; mycelia of fungi may live in the wood. In 1906 two fungi which had formed fruit-bodies were found on tree-trunks washed ashore. All these modes of transport may have contributed to make develop on the beaches of Krakatao a rich vegetation not exclusively composed of typical littoral plants. The lack of a mangrove-vegetation (consisting in the East Indian Archipelago of $\pm$ 20 species [1]) is due to the fact that on Krakatao there is no locality fit for it. Destruction by crabs of seeds washed ashore does not take place. The great number of coco-palms is remarkable.

The seeds of some littoral plants may have been carried over by birds either exozoically or endozoically. In the last way birds may have carried over at least 9 littoral species [2]). Though, with birds, the time between the consuming of food and the removal of the undigested rests of it from the intestine tract is not long (sometimes only $\frac{1}{2}-\frac{3}{4}$ hour, usually $1\frac{1}{2}-3$ hours), this is not to be considered an insurmountable obstacle for the transport of seeds, because the distance between Krakatao and the surrounding islands is but short. Also plants of the interior, *Ficus* for instance, may have been introduced by birds. For Verlaten Eiland, which was inhabited during some time [3]), introduction of a plant, *Carica Papaya L.* (a rapidly growing, in the Dutch Indies commonly cultivated fruit-tree) by birds is possible.

As it appeared during Treub's visit in 1886 the wind has been of great importance for the development of the new flora. Before the results of Treub's trip had been published most botanists were of opinion that the wind carried to remote islands only spores of ferns and lower Cryptogams but hardly any fruits or seeds. That, as a matter of fact, these latter must sometimes have been carried over was deduced from the presence of higher plants on coral islands. Some botanists however ascribe to the wind – especially in the higher regions of the air – a more important rôle though they also are of opinion that transport over long distances only very rarily takes place. The fact that in the East Indian Archipelago very remote mountain-tops bear the same species absent in the regions between, is ascribed by Beccari to the wind which, at least in the West-monsoon when it blows regularly, carries the often very small seeds from west to east [4]). According to the same author this transport can also have taken place for a part by birds. Against the first supposition of Beccari Ernst adduces the well-known statement of Junghuhn [5]) that above

[1]) The number given by Ernst is rather low; in reality it is more than double so large. Mangrove-components in the East Indian Archipelago are i.a: *Acanthus ebracteatus Vahl, ilicifolius L.* and *volubilis Wall.*; *Acrostichum aureum L.*; *Aegialitis annulata R. Br.*; *Aegiceras corniculatum Blanco* and *floridum R. et Sch.*; *Avicennia marina Vierh.* and *officinalis L.*; *Brownlowia argentata Kurz* and *lanceolata Bth.*; *Bruguiera caryophylloides Bl.*, *eriopetala W. et A.*, *gymnorrhiza Lamk.* and *parviflora W. et A.*; *Caesalpinia nuga Ait.*; *Cerbera manghas L.*; *Ceriops candolleana Arn.* and *roxburghiana Arn.*; *Clerodendron inerme Gärtn.*; *Cryptocoryne ciliata Fisch.*; *Cumingia philippinensis Vidal*; *Cynanchum carnosum Schlecht.*; *Cynometra ramiflora L.*; *Dalbergia menoeides Prain* and *torta Grah.*; *Derris heterophylla Backer*; *Diplachne polystachya Backer* (=D. fusca Auct.); *Dolichandrone spathacea Schum.*; *Excoecaria Agallocha L.*; *Fimbristylis ferruginea Vahl*; *Finlaysonia maritima (Bl.) Backer*; *Gymnanthera paludosa Schum.*; *Heritiera littoralis Dryand.*; *Ixora timoriensis Decsne*; *Kandelia candel Druce*; *Lumnitzera littorea Voigt* and *racemosa Willd.*; *Nephrolepis acutifolia Christ* (epiphytic); *Nipa fruticans Wurmb*; *Oncosperma tigillaria Ridl.*; *Paramignya littoralis Kurz*; *Phoenix paludosa Roxb.*; *Pithecolobium umbellatum Bth.*; *Rhizophora conjugata L.*, *mucronata Lamk.* and *stylosa Griff.*; *Sarcolobus Banksii R. et Sch.*; *Scyphiphora hydrophyllacea Gärtn.*; *Sesuvium portulacastrum L.*; *Sonneratia acida L. fil.*, *alba Smith* and *ovata Backer*; *Viscum orientale L.* (epiphytic parasite); *Xerochloa imberbis R. Br.*; *Xylocarpus granatum Koen.* and *moluccensis Lamk.*

[2]) To these nine Ernst (Neue Flora Vulkaninsel Krakatau (1907) p. 61) reckons also *Clerodendron inerme Gärtn.* to which he wrongly ascribes juicy fruits. The fruits of this plant have a thick dry wall and are spread by water.

[3]) This is an error. Not Verlaten Eiland but Lang Eiland (See p. 84 seq.) was inhabited during some time before 1906.

[4]) It is not clear how somewhat heavy seeds or fruits of low terrestrial herbs living in dense forests can be carried upwards by the wind which shakes the tops of the trees but does not reach the soil. Yet on very distant mountains the *same terrestrial plants occur in such forests*, f.i. *Coleus galeatus Bth.* – *Impatiens chonoceras Hassk.* – *Melissa parviflora Bth.* – *Ranunculus diffusus D. C.* The fruits or seeds of none of these plants possess either anemochorous or zoochorous adaptations. They are far too heavy to be carried upward (in the forest) by ascending air-currents.

[5]) See Junghuhn, Java, 2nd Dutch Edition I, 225; German Edition I, 166. This statement has since proved incorrect. See Braak, Het Klimaat van Nederlandsch Indië p. 11 in ann., pp. 75. 76, 77.

an altitude of 2000 m.[1]) the south-east trade-wind blows throughout the year. Vagler has proven for the Swiss Alps that there during storms a transport of seeds over a distance of 20 km. is possible. In 1897 on Mount St. Gotthard a hail of salt-crystals fell, the biggest of which had a weight of 4 g.; these crystals must have been transported over a distance of at least 250 km., perhaps much more. The best proof, however, that plant-germs may be carried through the air over large distances is furnished by Krakatao; the seeds carried thither by the wind must originate from islands lying at distances of 18 km. or more. The first vegetation of the interior of Krakatao arose almost exclusively from plant-germs introduced by the wind. These plants were algae, bacteria, diatoms, Hepaticae, Musci, ferns and at least 6 of the 8 Phanerogams found by Treub in the interior of the island. Ferns predominated with 11 species. According to Penzig, till 1897 17 species, 32% of the total flora, were brought over by the wind. — Ernst believes that till 1906 28 species, 30% of all Phanerogams may have been carried to Krakatao by the wind; for a number of these — especially plants growing on or near the beach — he considers transport by water also as possible. Ernst considers the arguments adduced by Schimper against transport of seeds by wind across the sea as invalid. Light seeds, even those that possess no special adaptations for floating through the air, can be spread over a large distance by winds having a velocity of 30 m. per second According to Ernst such winds are in the East Indian Archipelago during storms by no means of rare occurrence. Hence light seeds and fruits not possessing adaptions for floating through the air may have been carried across the Sunda Straits, though such may more easily happen with seeds provided with such adaptions. Spores and very light seeds, like those of orchids, may have been transported by ordinary trade-winds; the heavier seeds — *if* at least they were transported by the wind — by heavy storms.

From the above appears that we cannot exactly tell which part each of the 3 agents of dispersal has taken in the revegetation of Krakatao. Probably the orchids were brought by the wind, though another mode of transport is not to be considered impossible; neither need the other anemochorous species have been imported by the wind. The seeds and fruits of beach-plants may have come over either floating on the sea or in crevices and hollows of driftwood and pumice; they may also have been carried over by birds. Of the plants found till 1906 on Krakatao at least 39%, perhaps even 67—72%, were brought there by sea-currents. — Nor can we make out with certainty how many plants were imported by birds. Nearly all plants found on Krakatao have a wide distribution over the Malayan Archipelago. The flora of the regions adjacent to the Sunda Straits being very imperfectly known, we cannot trace the origin of such species found on Krakatao as are not common in the Dutch Indies. Recapitulating, Ernst arrives at the conclusion that by the wind were imported nearly all Cryptogams and 16—30% of the Phanerogams, by sea-currents 39—72% of these latter, the rest by birds. The import of so many Cryptogams should be ascribed to the fact that the winds blowing in the Sunda Straits are strong enough to carry light spores over the short distance that separates Krakatao from the surrounding islands. On the much more remote Cocos-islands ferns are lacking[2]). Only 3 Cryptogams (a fungus, a moss and a Lichen) have been found there.

To this admirably lucid statement I need add but very little. Doubtless plants of the interior *may* have been imported in various ways, but of not a single species found till 1906 we *know* with certainty that it has been imported, nor in which manner. That fruits

[1]) Junghuhn says 2000 (old French) feet (of 0,3248395 m.) = 1950 m.

[2]) That this *lack* of ferns is indeed to be ascribed to the long distance separating the Cocos-islands from Java and not to climatic or edaphic influences, is no more proven than the precisely opposite statement of Treub that the long distance separating Juan Fernandez and Ascension from the nearest continent is the cause of the *abundance* of ferns in these islands. See p. 76 seq.

and seeds are spread by wind, water and animals was known already long before the eruption. But it was expected that from the investigations on Krakatao would appear *that* and *in which manner* seeds and fruits are transported across a sea-arm by other means than sea-currents. That expectation has *not been fulfilled.* Now that the total destruction of the vegetation in 1883 has not been proven and not a single experiment has been made in order to ascertain whether seeds or fruits really were carried across by the wind or by animals, one can only state that the trips to Krakatao have contributed *nothing, litterally nothing,* to the advancement of our *real knowledge* – not our *fantasy* – in this respect. Visitors have found plants, they have *assumed* that these were imported after the eruption and then they have set themselves to *guessing.* Light seeds they have had introduced by the wind, fleshy fruits by birds, everything according to the old recipe. But *nothing* has been *proven,* none of the so-called results has a scientific base.

In the last chapter of his paper Ernst traces in bold outlines the development of the flora of Krakatao.

In 1897 on the beach a *Pes-caprae* formation was found, in the interior a grass-steppe, on the mountain slopes a fern-vegetation. In 1906 the situation had very much changed. On the beach not only a *Pes-caprae*-formation but also a (not yet very rich) *Barringtonia*-formation occurred; littoral plants were growing to 300–500 m. from the sea; conversely, arms of the grass-steppe which covered the interior projected through gaps in the littoral forest as far as the beach. In the south-eastern [1]) angle of the island the littoral vegetation consisted of groups of different ages, the younger of these were found close to the sea, the older ones more towards the interior; this points to the beach having shifted. Considering the composition of the fauna of Krakatao in 1906 it is not probable that seeds of littoral plants were carried by animals to the interior. The gentle slopes behind the littoral forest and the lower vallies and ridges were in 1906, like in 1897, clothed with grasses, *Cyperaceae* and *Compositae*; ferns were already in the minority. Trees and shrubs were in 1906 more numerous than in 1886. In stray specimens or in groups they had penetrated from the littoral forest into the interior; in the ravines they had blended into groves.

If no new eruption takes place, the entire island, except the steep wall formed by the rupture, will be covered after 50 or 60 years with forest. The littoral groves will blend into each other, new species will appear, others will disappear or retreat to the *Pes-caprae*-formation and locally supersede it. A number of the facultative beach plants will be carried to the interior by the wind or by animals. Possibly the forest in the higher ravines consisted in 1906 already of these littoral plants; subsequent visitors may find there numerous other species brought there by birds and yet more by the wind from other islands, especially from the higher regions of these. The south-east trade-wind can bring many plants from Java. Especially those anemochorous and zoochorous species may be expected which in Java and Sumatra form a new flora on volcano-cones deprived by an eruption of their vegetation. As the woods expand and grow denser new conditions arise enabling other plants to live there. The influence of man on Krakatao will be very feeble, the woods will continually extend themselves, the grass-steppe will disappear, but not before many years have elapsed the vegetation will be as rich as that of a tropical primeval forest.

[1]) By some lapsus calami Ernst wrote south-*western* angle.

With these views I can only partly agree. It is not *proven* that in former years the entire south-eastern part of the island bore a fern-vegetation like that which Treub and Penzig found on the basaltic rocks of Zwarte Hoek. It is highly improbable that the basaltic rocks will ever bear a forest, that littoral plants ever ran high up against the mountain, that the young forest of the interior has originated from the littoral one. The trees found at some distance from the beach belonged in 1906 exclusively to such species as are never found on the beach proper and those parts of the ravine-forest between 200 and 400 m. above sea-level which were reached in 1908, did not contain a single littoral species. On p. 26 I argued already that in my opinion it were the higher ravines which formed an important starting-point for part of the new vegetation. Of course I will by no means deny that introduction from outside the island might in 1906 already have taken place with plants of the interior, but all data required for drawing a safe conclusion on this point are lacking.

IX. THE EXPEDITION OF FRANSEN HERDERSCHEE, HOLTAPPEL, COOL, JACOBSON, DEMMENI AND BACKER IN MAY 1908.

Jaarverslag van den Topographischen Dienst in Ned. Indië (Year-book of the Topographical Survey Service in the Dutch Indies) 1908, p. 150.

In May 1908 a topographical and geological survey was made of the three islands of the Krakatao-group. The topographical observations were made by A. Fransen Herderschee and H. W. Holtappel, accompanied by the photographer J. Demmeni and some native underlings. Mr. E. R. Jacobson accompanied the expedition in the quality of a zoologist. Shortly before the departure of the expedition Prof. Treub ordered me to join it for the purpose of making botanical collections. As I had neither the time nor the experience necessary for procuring the outfit and the food required for camping on an uninhabited island I could take part only in the beginning of the trip but had to leave before the top was reached. The experiences of the expeditionists are recorded in the paper mentioned above.

We were conveyed to Krakatao by the small steamer Java which left Tandjong Priok on May 2nd 1908 and in the morning of the following day reached Telok Betong (South Sumatra) where some officials of the Topographical Survey Service embarked. Part of these were taken to different islands in the Sunda Straits, after which the vessel made for the south-eastern angle of Krakatao, about the same point where the excursionists of 1906 had landed. Messrs. Fransen Herderschee, Holtappel and Jacobson debarked there on May 4th. We arranged that they should try to cut a path from that point to the top whilst I myself should stay on the steamer in order to visit consecutively Zwarte Hoek, Verlaten Eiland and Lang Eiland. I should return on May 6th to the south-eastern angle of Krakatao from where I should ascend the mountain along the path made in the mean time by the 3 others. Little did we foresee the difficulties that were in store for us!

Shortly after noon on May 4th the Java dropped anchor off Zwarte Hoek. When steaming past the steep wall formed by the rupture I could observe that in many spots it bore living plants. From 3 till 5 in the afternoon I had the occasion to herborize on this beach, explored already by so many botanists, and on the base of the basaltic rocks. The following 52 species were found:

Polypodiaceae.

1. *Nephrolepis exaltata Schott.* See p. 56, No. 6.
   Rather numerous on the older parts of the beach, in groups.
2. *Ceropteris calomelanos Und.* See p. 54, No. 1.
   Very numerous on the lower parts of the steep wall formed by the rupture.
3. *Drynaria quercifolia J. Sm.* See p. 113, No. 26.
   One specimen in a crevice at the base of the basaltic rocks.

Lycopodiaceae.

4. *Lycopodium cernuum L.* See p. 113, No. 29.
   A few specimens at the base of the basaltic rocks.

Pandanaceae.

5. *Pandanus tectorius Sol.* See p. 43, under No. 4.
   Several specimens as well on the younger parts of the beach as on the somewhat older ones.

Gramineae.

6. *Imperata cylindrica P. B.* See p. 106.
   Some specimens in crevices at the base of the basaltic rocks.
7. *Saccharum spontaneum L.* See p. 100.
   Very numerous on the older parts of the littoral plain and on the tali formed by ashes sliding down.
8. *Pogonatherum paniceum Hack.* See p. 105.
   Very numerous on the steep wall formed by the rupture and on the tali at the foot of it.
9. *Ischaemum muticum L.* See pp. 89, 90.
   Rather numerous on the beach.
10. *Pennisetum macrostachyum Brongn.* See p. 48, No. 8 and p. 99.
    As well on the base of the basaltic rocks as on the leached-out older parts of the littoral plain, nowhere numerous. It seemed to be less common than in 1906.
11. *Spinifex littoreus Merr.* See pp. 89, 90.
    Very numerous on the beach, seemingly more common than in 1906.

Cyperaceae.

12. Cyperus pennatus Lmk. See p. 47, sub No. 5 and 6 and p. 90.
Several specimens on the older firmer parts of the littoral plain.

13. Fimbristylis spathacea Roth. See p. 47, sub No. 5 and 6 and p. 90.
Numerous on the older firmer parts of the beach.

Palmae.

14. Cocos nucifera L. See p. 39.
Some very young, probably planted specimens rather far from the flood-mark, 300 à 400 m. to the south-east of Zwarte Hoek. The specimens seen two years before were not found back; they must have died by some cause or other or have been removed.

Orchidaceae.

15. Spathoglottis plicata Bl. See p. 107.
Some specimens on the wall formed by the rupture.

16. Eulophia macrorrhiza Bl.
Saprophytic leafless orchid with a large subterraneous or partly superterraneous tuber, found here and there in West-Java, from the plains up to an altitude of $\pm$ 1000 m. mostly on calcigerous soils and on brecciae, rarely on lateritic formations. Though, as a rule, a rare species it may be locally rather common. The reproduction is effected by the seeds.
Three specimens on the older parts of the littoral plain.

Casuarinaceae.

17. Casuarina equisetifolia Forst. See p. 91, No. 6.
A few rather young specimens.

Ulmaceae.

18. Trema orientale Bl. See p. 132, No. 15.
Several specimens on the older parts of the littoral plain.

Moraceae.

19. Ficus fulva Reinw. See p. 133, No. 17.
Several specimens on the older parts of the littoral plain.

Urticaceae.

20. *Pouzolzia zeylanica Benn.*

Halophobous perennial of sunny or slightly shadowed, humid or marshy soils, spread throughout Java from the plains to an altitude of 1600 m., often occurring on water-sides. The fruits remain enclosed within the (after the anthesis) somewhat enlarged perianth, hence possess floating power.

A few specimens on the base of the wall formed by the rupture.

Olacaceae.

21. *Ximenia americana L.*

Spinous shrub, found in Java in many places but exclusively on the beach. The reproduction is effected by the fruits [1]) which possess floating power because of the testa being thick and corky.

One specimen on the beach.

Lauraceae.

22. *Cassytha filiformis L.* See p. 91, No. 7.

Numerous, parasitizing on all kinds of plants, as well on the beach as on the older parts of the plain.

Leguminosae.

23. *Pithecolobium umbellatum Bth.* See p. 147, No. 20.

Two specimens on the littoral plain.

24. *Albizzia retusa Bth.*

Small tree, found in Java here and there on sandy beaches rich of lime. The pods which possess a rather thick corky rind do not burst when ripe but easily break into fragments each of which contains one or more seeds. These fragments possess floating power.

One young specimen on the beach.

25. *Sophora tomentosa L.* See p. 147, No. 21.

One specimen somewhat behind the floodmark.

26. *Desmodium umbellatum D. C.* See p. 148, No. 22.

Here and there on the littoral plain.

---

1) Described by Schimper, Indo-malayische Strandflora (1891), p. 176.

27. *Derris heterophylla Backer.* See p. 134, No. 21.
A few scattered specimens on the younger parts of the beach.

28. *Erythrina fusca Lour.* See p. 46, under No. 1.
One young specimen behind the flood-mark.

29. *Erythrina variegata L.* var. *orientalis Merr.* See p. 46, under No. 1.
Scattered young specimens on the littoral plain.

30. *Canavalia rosea D. C.* See p. 89 and p. 92, No. 8.
Numerous on the littoral plain, apparently more common than in 1906.

31. *Vigna marina Merr.* See p. 89 and p. 93, No. 9.
Numerous on the littoral plain.

Euphorbiaceae.

32. *Euphorbia atoto Forst.* See p. 90 and p. 93, No. 10.
One specimen on the beach.

Sapindaceae.

33. *Dodonaea viscosa Jacq.* See p. 148, No. 28.
Some specimens of the glabrous shrubby littoral form on the younger parts of the beach.

Malvaceae.

34. *Hibiscus tiliaceus L.* See p. 93, No. 11.
A few young specimens on the littoral plain.

Guttiferae.

35. *Calophyllum inophyllum L.* See p. 44, No. 6.
Several young specimens on the beach.

Lecythidaceae.

36. *Barringtonia asiatica Kurz.* See p. 43, No. 5.
A few very young specimens on young parts of the beach.

Combretaceae.

37. *Terminalia catappa L.* See p. 39 No. 2.
Rather numerous on the younger parts of the littoral plain.

Melastomataceae.

38. *Melastoma malabathricum L.* See p. 150, No. 33.
Several specimens on the base of the basaltic rocks.

Convolvulaceae.

39. *Ipomoea pes caprae Sw.* See p. 48, No, 7.
Very numerous on the littoral plain where it, locally, formed a denser covering than in 1906.

Verbenaceae.

40. *Premna integrifolia L.* See p. 137, No. 37.
Scattered on the littoral plain.

41. *Vitex paniculata Lamk.*
Shrubby species often — though in my opinion wrongly so — considered as a mere synonym of *Vitex Negundo L.*, occurring throughout Java on sandy beaches The reproduction is effected by the drupes, the kernels of which possess floating power.
A few scattered specimens on the beach.

42. *Clerodendron inerme Gärtn.* See p. 151, No. 37.
Here and there on the littoral plain.

Labiatae.

43. *Hyptis brevipes Poit.* See p. 138, No. 38
Numerous specimens on the base of the basaltic rocks.

Rubiaceae.

44. *Oldenlandia corymbosa L.*
Small annual, spread throughout Java from the plains to an altitude of ± 800 m, in sunny not too dry localities, especially on hard or stony soils, along road-sides and dikes, in premises and in fields, locally often common. The reproduction is effected by the very small seeds which are produced in great numbers and possess floating power.
Some specimens on the base of the basaltic rocks, in crevices.

45. *Morinda citrifolia L.* See p. 93. No. 16.
Several specimens on the littoral plain.

Goodeniaceae.

46. *Scaevola frutescens Krause.* See p. 49, No. 9.
Rather numerous on the beach also on its older parts.

Compositae.

47. *Vernonia cinerea Less.* See p. 118, No. 58.
Rather numerous on the foot of the basaltic rocks and on the older parts of the littoral plain.

48. *Erigeron linifolius Willd.* See p. 51.
A few specimens at the base of the steep wall which secludes the littoral plain from the interior.

49. *Blumea lacera D. C.* See p. 51.
Several specimens on and near the base of the basaltic rocks.

50. *Erechthites hieracifolia Rafin.* See p. 52.
Some specimens on the base of the basaltic rocks.

51. *Pluchea indica Less.* See pp. 50, 108 and 119.
Here and there on the older parts of the littoral plain.

52. *Wedelia biflora D. C.* See p. 49 under *Wollastonia.*
Numerous, especially on the older parts of the beach.

From a comparison of the list above with that of the plants collected in the same locality by Ernst and his companions in 1906, it appears that the vegetation in the two intervening years had but little changed. In 1908 were not found back *Dryopteris unita O. K.* (in 1906 a few specimens on the base of the steep wall formed by the rupture) and *Cerbera manghas L.* (in 1906 a few very young seedlings on the younger parts of the beach). Whether the first of these two species was really absent in 1908 or was overlooked in the locally dense vegetation I cannot say; it is quite possible that the few specimens growing there in 1906 were all taken away by the botanists then visiting the island. The young seedlings of *Cerbera manghas L.* collected in April 1906, were found in a locality unfit for their further growth, a beach consisting of very loose sands (see p. 45); they probably died in the east-monsoon of that year. In the westmonsoon very frequently seeds of all kinds germinate on such beaches but do not find a fit habitat there and perish in the course of the next east-monsoon.

Found in 1908 but not in 1906 were 8 species, namely *Spathoglottis plicata Bl.* (a few specimens), *Eulophia macrorrhiza Bl.* (3 specimens), *Pouzolzia zeylanica Benn.* (a few specimens), *Ximenia americana L.* (1 young specimen), *Albizzia retusa Bth.* (1 young specimen), *Erythrina fusca Lour.* (1 young specimen), *Vitex paniculata*

Lamk. (a few scattered specimens) and *Oldenlandia corymbosa L.* (a few specimens). All these species were represented by so very few specimens that they could not influence the character of the vegetation. It cannot be decided whether these „new species" were imported after 1906 in the investigated locality or already occurred before that time; a few young specimens of a species or other may quite well have been present in 1906 but have been overlooked. When *Eulophia macrorrhiza* is in a period of rest and bears no flowers it is very inconspicuous or quite invisible. The species of *Pouzolzia, Ximenia, Albizzia, Erythrina* and *Vitex,* found at Zwarte Hoek in 1908, have fruits fit to be spread by water, they *may* have been carried to Zwarte Hoek after the trip of 1906. For *Ximenia, Albizzia* and *Erythrina* which were represented only by young specimens, this is not improbable. It cannot at all be decided whether these species had come over from another point of the island or were introduced from without. The two orchids (*Spathoglottis plicata Bl.* and *Eulophia macrorrhiza Bl.*) possess very minute seeds wich can easily be spread over some distance by the wind or by rain-water floating over the ground, but it is not al all certain that they had come from outside the island. *Spathoglottis* was already found in 1897 on Krakatao in numerous specimens, *Eulophia* was found for the first time in 1908, but may have grown on Krakatao long before that year, even before the eruption. All investigations have been carried out so hastily and over such a small part of the island, that nobody can say with certainty that a species in a given year was *not* present. *Oldenlandia corymbosa* also can have grown long before the eruption on the rocks at Zwarte Hoek, its seeds may quite well have survived the eruption. From the finds at Zwarte Hoek in 1908 we may perhaps conclude that the flora of this locality had grown somewhat richer since 1906, but this conclusion may not be extended to the entire island. The little modifications in the frequency of a few species (Nos. 10, 11, 30 of the list here before) are of no importance.

On May 5th I made a trip in the northern and north-eastern parts of Verlaten Eiland, on May 6th a trip on the north-eastern and eastern parts of Lang Eiland. Of both islands only the lower parts were visited, where a dense and rather rich vegetation (resp. 64 and 64 species) was found, which I shall not discuss here (See pp. 34 and 35).

In the evening of May 6th I went back to the south-eastern point of Krakatao, where I hoped that my companions, whom I had left two days before, would have succeeded in the meanwhile in clearing a path to the top. But the many serious difficulties offered by the locality and by the vegetation had caused that their efforts had not been crowned with success. An excellent report of their doings was given by Fransen Herderschee in the paper cited at the head of this chapter. From this paper I derive the following data:

The lower slopes on the south-eastern side of the island were still clothed with a very dense vegetation of *Saccharum spontaneum L.* (glagah). When after a toil of many hours a number of coolies had cut a path through that vegetation the explorers made on May 5th the first effort to reach the top of the mountain. After having crossed with much trouble some narrow steep-sided ravines they reached a broad, less deep ravine the stony bottom of which bore a meagre vegetation; this ravine seemed to be an old lava-stream; it did not run through but branched into several smaller ones. On the branching point the explorers ascended one of the sides of the ravine and continued their way along one of the ridges, which was repeatedly traversed by narrow crevices. After a weary march of some hours they had made but little progress. Therefore they descended into the nearest ravine of 5—6 m. depth, the bottom of which consisted of volcanic ashes. The vegetation of this ravine (See p. 187) consisted mainly of trees of 10—15 m. height and of three species of terrestrial orchids. In an upward direction the ravine continually narrowed till it ended in a narrow cleft. Then the explorers cut their way over a narrow *Saccharum*-clad ridge till about 50 m. further a new ravine was reached. In the mean time it had become noon, for which reason they decided to arrange a camping-place for the night on the side of the ravine. Most of the coolies they sent back to the beach to fetch food, water and other necessaries; the trip was continued with 2 coolies only. After excessive efforts a main-ravine was reached at least in which the original rock-formation, the old volcano-mantle, lay bare. Then it had become too late to go on and the explorers returned to the camping-place.

In the early morning of May 6th the trip was continued. Soon already the explorers reached the main-ravine in which they made considerable progress. Here the trees were higher (10—20 m.) than in the lower ravines and gave rather much shadow; ferns and grasses were also found. Suddenly the way was blocked by large rocks behind which, however, the ravine continued. The ascent became continually more difficult; at every moment rocks obstructed the way. At an altitude of ± 275 m. the explorers passed a steep, slippery cataract (then waterless) of ± 5 m. height which was clothed with grasses and ferns (see p. 188). Above this cataract the ravine divided into branches of which the explorers followed the western-most one, bounded on the east-side by a vertical wall. After having ascended somewhat less than 100 m. the explorers were arrested by a vertical wall which they climbed with the aid of a tree-trunk. By doing so they reached a ridge not covered with *Saccharum* but mainly with shrubs. Towards the top of the mountain this ridge became gradually narrower; at least it had a breadth of only $1/2$ m. and was bounded on either side by a deep ravine. At an altitude of 400 m. the ridge was suddenly interrupted by a deep transverse cleft. As it had become too late to go on the explorers decided to return to the camping-place in the ravine.

So far the report of Fransen Herderschee. On May 7th I myself, in the company of the others ascended the south-eastern slopes thereby following mostly the path cut under the direction of

F r a n s e n H e r d e r s c h e e. The officials of the Topographical Survey Service having found for the lower part of the path a better course we were able to reach the camping-place from the south-eastern point of the island within an hour. From there we continued our way as far as the end of the path made the previous day; it was as yet impossible to go further.

The littoral vegetation [1]) and the *Casuarina*-forest had quite the same aspect as two years before (See p. 152, seq.), therefore I shall refrain from describing them. The tallest *Casuarina* had — roughly measured — a height of ± 35 m. The coco-palm grove (See p. 177, No. 11) was still present and had much the same appearance as in 1906. Behind the littoral vegetation here and there a very narrow strip of mixed forest was lying, consisting of 5 — 15 m. high specimens of *Ficus fistulosa Reinw.*, *Ficus fulva Reinw.*, *Macaranga tanarius M. A.*, *Melochia umbellata Stapf* and *Pipturus incanus Wedd.*, often partly overgrown with *Trichosanthes bracteata Voigt* and *Columella trifolia Merr.* — Immediately behind this mixed forest or, wherever it was absent, behind the littoral vegetation, the broad and very dense girdle of *Saccharum spontaneum L.* began, which showed, however, rather many scattered small hiatuses in which besides the 5 trees named above (especially *Pipturus incanus Wedd.*) a number of herbaceous kremnophytes occurred, mostly in poor specimens viz. *Arundina speciosa Bl.* which was rather numerous, *Spathoglottis plicata Bl.* which occurred in scattered specimens and a few ferns as *Dryopteris unita O. K.* and *Dryopteris setigera O. K.* On the steep walls of eroded gullies *Lycopodium cernuum L.* was rather frequent.

The *Saccharum*-girdle had a breadth of several hundreds of meters. We traversed it rapidly because we wished to give as much of our time as possible to the investigation of the forest in the ravines above it. This in many places very dense forest consisted in the ravines explored by us of very few species, often represented by innumerable individuals, mainly of *Ficus fistulosa Reinw.*, which reached in the lower ravines a height of 10 — 15 m, in the heigher ones of 15 — 20 m. Other components of the ravine-forests were *Macaranga tanarius M. A.* (not rare), *Ficus fulva Reinw.*, *Ficus toxicaria L.*, *Homalanthus populnea O. K.* (upwards of 200 m. here

[1]) The only species found in 1908 on the beach which had not yet been found in 1906 was *Luffa cylindrica Roem.* (1 specimen), a Cucurbitacea of which a small-fruited wild form not seldom occurs on water-sides or on the beach. A form with large fruits is often cultivated by the natives in the interior.

and there) and *Pipturus incanus Wedd.* (rather numerous). Several species of herbs and shrubs formed the undergrowth, namely *Costus sericeus Bl.* (upwards of 100 m. numerous), *Leea indica Merr.* (upwards of 100 m, here and there), – *Cyrtandra sulcata Bl.* (especially upwards of 250 m. very numerous), – *Leucosyke capitellata Wedd.* (rather numerous above 300 m), – *Heckeria umbellata Kth.* var. *subpeltata D. C.* (numerous above 200 m), – *Nauclea purpurascens Korth.* (one specimen at 330 m), – *Phajus tankervilliae Bl.* (here and there upwards of 200 m) and *Selaginella plana Hieron.* (numerous upwards of 200 m). On the bottom of the ravines here and there a small terrestrial orchid, *Nervilia aragoana Gaud.*, in its leaf-bearing state (see the list hereafter sub No. 13), was found. The steep walls of the ravines bore numerous kremnophytes, especially *Dryopteris unita O. K.* and *Dr. setigera O. K.*, – *Pogonatherum paniceum Hack.*, – *Spathoglottis plicata Bl.* (which covered many ravine-sides from above to beneath) and *Arundina speciosa Bl.*, moreover (between 250 and 300 m.) scattered specimens of *Ficus quercifolia Roxb.* On and near the slippery cataract mentioned above several groups of *Nephrolepis exaltata Schott* and *Pogonatherum paniceum Hack.* were growing.

On the ridges, besides the predominant *Saccharum spontaneum L.*, *Pennisetum macrostachyum Brongn.* was found up to an altitude of 330 m. in scattered specimens or in small groups but never gregariously.

Epiphytic ferns were rare, in Java a very common phenomenon in young forest. Only three species were found, viz. *Polypodium phymatodes L.*, – *Polypodium punctatum Sw.*, both very common ferns in the lower regions of Java and *Drynaria quercifolia J. Sm.* This last species occurred also as a kremnophyte on steep walls.

The day after this ascent which was only a very poor success or rather a failure – the most important parts of the mountain, the higher ravines not having been reached – our steamer had to go back to Tandjong Priok. As I mentioned already above I had not brought the outfit and the food necessary for a prolonged stay; therefore I was reluctantly obliged to embark and to return to Buitenzorg.

Two days afterwards Fransen Herderschee and Demmeni made a new effort to reach the top. At a height of 300 m. they turned to the east. Under a fore-top they saw three high tree-ferns above which the slopes seemed fit for further ascent. Therefore they continued their way in the direction of these ferns. Every moment they had to cross small ravines; the *Saccharum*-vegetation was less dense and less high and was gradually replaced by ferns, whilst here and there a specimen of *Ficus* was growing. At two o'clock in the afternoon a height of 475 m. was reached, where the explorers made their camp for the night.

The following day, May 10th, Demmeni alone continued the trip and succeeded at last in reaching the top. The 3 tree-ferns mentioned above, the only ones found, were growing at an altitude of about 600 m. On the top there were no trees but herbaceous ferns, intermixed with a few shrubs and with *Saccharum*.

Here beneath I give a complete survey of the plants observed by me in 1908 in the interior.

Polypodiaceae.

1. *Dryopteris setigera* O. *K.* See p. 57, under No. 8.

Scattered in open places in the *Saccharum*-girdle. Numerous on the steep walls of the gullies and ravines above an altitude of 50 m.

2. *Dryopteris unita* O. *K.* See p. 56, under No. 7,

Scattered in open places in the *Saccharum*-girdle. Numerous on the steep walls of the gullies and ravines above an altitude of 50 m.

3. *Nephrolepis exaltata Schott.* See p. 56, No. 6.

Several groups at an altitude of ± 275 m. on and near a humid stony cataract.

4. *Polypodium phymatodes L.*

Epiphytic or less often terrestrial fern, spread throughout Java from the plains till an altitude of 1500 m., on the whole rather common, but absent or scarce in many regions where the east-monsoon is very strong. It prefers a moderately shadowed habitat. The reproduction is effected by spores and by ramification of the rather long rhizome.

Some scattered epiphytical specimens in ravines at an altitude of 200 m.

5. *Polypodium punctatum Sw.*

Epiphytic fern spread from West- to East-Java, from the plains to an altitude of ± 1500 m., except in regions where the east-monsoon is very strong. The reproduction is exclusively effected by spores.

Some scattered epiphytical specimens in ravines at an altitude of 200 m.

6. *Drynaria quercifolia J. Sm.* See p. 113, No. 26.
Some scattered specimens, partly living as epiphytes on trees, partly as kremnophytes on ravine-sides.

Lycopodiaceae.

7. *Lycopodium cernuum L.* See p. 113, No. 29.
Numerous on the steep walls of the lower ravines.

Selaginellaceae.

8. *Selaginella plana Hieron.*
Erect or, at the very base, decumbent herb, spread throughout Java from the plains up to an altitude of ± 750 m. except in regions where the east-monsoon is very strong. It is found in shadowed humid mostly not very fertile localities, on steep earth-walls, in thickets and secondary forests, locally often in great numbers.
Numerous in the ravines between 200 and 400 m. above sea-level.

Gramineae.

9. *Saccharum spontaneum L.* See p. 100 seq..
Very numerous from just behind the beach to an altitude of 400 m., beneath 50 m. forming a very dense girdle, above that altitude occurring especially on the ridges. Found by Demmeni up to the summit (813 m.).

10. *Pogonatherum paniceum Hack.* See p. 105.
Numerous on steep sides of ravines and gullies.

11. *Pennisetum macrostachyum Brongn.* See p. 48, No. 8, and p. 99.
From the lower parts up to an altitude of 330 m. on the ridges in scattered specimens or in small groups, on the whole rather rare.

Zingiberaceae.

12. *Costus sericeus Bl.*
Robust herb with dehiscent capsules, spread throughout Java from the plains to the lower mountainous zone, as well in regions where the east-monsoon is rather strong as where

it is feeble, in thickets and secondary forests. The reproduction is effected by the seeds which are provided with a rather soft fleshy black testa and an incomplete white arillus. It seems that, sometimes at least, the seeds are sought after by animals for I repeatedly saw seedlings growing in the pouches formed by the persistent bases of the leaf-stalks of oil-palms (*Elaeis guineensis Jacq.*) and gomuti-palms (*Arenga pinnata Merr.*), rather high up in the trees.

Orchidaceae.

13. Nervilia aragoana Gaud.

Small terrestrial orchid spread throughout Java between 30 and 1000 m. above sea-level, especially on periodically dry heavy soils in shadowed localities, often in forests of teak (*Tectona grandis L.f.*) and of the Javanese thorny bamboo (*Bambusa spinosa Bl.*), frequently growing gregariously. The plant possesses in the soil a small tuber which sends forth alternately, with periods of rest between, a broad leaf and an inflorescence. It flowers in Java from September till February. The reproduction is effected by seeds.

A few stray leaf-bearing specimens in the lower parts of the ravine-forest.

14. Arundina speciosa Bl. See p. 107.

Not rare on the lower slopes, especially on open spots in the *Saccharum*-vegetation, always in scattered specimens.

15. Phajus tankervilliae Bl. See p. 107.

Here and there in the lighter parts of the ravine-forest above 200 m.

16 Spathoglottis plicata Bl. See p. 107.

Scattered in the lower parts of the island in open spots in the *Saccharum*-jungle. Above an altitude of 90 m. much more numerous, clothing entire ravine-walls.

Piperaceae.

17. Heckeria umbellata Kth. var. *subpeltata D. C.* (= Heckeria subpeltata Willd).

Robust herb, found in the western part of West-Java and in many places in East-Java between 125 and 1000 m. above sea-level in shadowed, constantly humid localities, especially in ravines and along watercourses, locally often rather numerous.

The reproduction is effected by the minute berries which are copiously produced. It is quite unknown in which manner the seeds are dispersed. The statement of Mr. Docters van Leeuwen that this is done by animals [1]) *may* be correct but probably flowing water also does its part.

Numerous in the ravine-forest above 200 m.

Moraceae.

18. Ficus fistulosa Reinw. See p. 133, No. 16.

From just behind the beach up to an altitude of 400 m. Numerous in the ravines above 200 m. where it was the main-component of the forest.

19. Ficus fulva Reinw. See p. 133, No. 17.

From just behind the beach to an altitude of 400 m. in scattered specimens or small groups.

20. Ficus quercifolia Roxb.

Small shrub, spread from the plains to an altitude of ± 1300 m. It grows in sunny or moderately shadowed, unfertile as well as fertile localities, often on steep walls, moreover on forest borders and in not too dark woods, frequently in teak-woods.

Here and there on ravine-sides between 150 and 300 m. above sea-level.

21. Ficus toxicaria L. fil. See p. 115, No. 44.

Here and there in the ravines between 200 and 400 m. above sea-level.

Urticaceae.

22. Pipturus incanus Wedd. See p. 134, No. 19.

In open places in the *Saccharum*-girdle and in the ravine-forest, rather numerous but mostly in scattered specimens.

23. Leucosyke capitellata Wedd.

Shrub, spread from West- to East-Java between 200 and 2400 m. above sea-level, mostly above 600 m. in shadowy forests, especially on fertile volcanic soils. The reproduction is effected by the somewhat fleshy fruits.

Rather numerous in the ravine forest from an altitude of 300 m. upwards.

---

1) Handelingen Eerste Nederlandsch-Indisch Natuurw. Congres Weltevreden (1920), p. 60.— Ann. Jard. Bot. Buitenzorg XXXI (1921), p. 129.

Euphorbiaceae.

24. *Macaranga tanarius M. A.* See p. 135, No. 26.
Scattered on the lower slopes. Not rare in the ravine-forest.

25. *Homalanthus populnea O. K.*
Small tree, spread throughout Java between 100 and 1800 m. above sea-level, especially in the western part of the island. It grows in sunny or slightly shadowed localities, in forest-glades and on forest-borders, on ravine-sides, in secondary forests and in jungles, frequently on water-sides, often in great numbers. The reproduction is effected by the seeds which are copiously produced and are dispersed i.a. by water. Though it seems not yet to have been observed they are possibly also spread by animals; if not, the occurrence of the tree far from water could hardly be explained, as the seeds are too heavy to be carried by the wind over a distance worth mentioning.
Here and there in the ravine-forest upwards of 200 m. above sea-level.

Vitaceae.

26. *Columella trifolia Merr.* See p. 136, No. 28.
Several specimens somewhat behind the beach.

27. *Leea indica Merr.*
Erect shrub or small tree spread throughout Java in the plains and in the lower mountainous zone, as well in regions where the east-monsoon is strong as where it is feeble, in thickets and secondary and primeval forests. The reproduction is effected by the berries which are produced in great numbers. It is unknown in which manner these are dispersed.
Here and there scattered in the ravine-forest above an altitude of 100 m.

Sterculiaceae.

28. *Melochia umbellata Stapf.* See p. 117, No. 51.
Scattered specimens immediately behind the littoral vegetation and here and there in the *Saccharum*-jungle.

Gesneriaceae.

29. *Cyrtandra sulcata Bl.*

Robust shrub spread throughout Java between 800 and 1400 m. above sea-level in shadowy humid forests. The reproduction is effected by the berries which contain numerous minute seeds. Whether animals take part in dispersing these seeds is unknown.

Numerous in the ravines above an altitude of 200 m.

Rubiaceae.

30. *Nauclea purpurascens Korth.*

Possibly the same species as was found by V a l e t o n in 1905, of which no materials were preserved. *Nauclea purpurascens* is a small or medium-sized tree, spread throughout Java from the plains up to an altitude of 1200 m. in shadowy secondary or primary forests and on river-sides, especially in regions where the east-monsoon is feeble. The reproduction is effected by the winged seeds which are produced in great numbers.

One specimen was seen in the ravine-forest at an altitude of ± 330 m.

Cucurbitaceae.

31. *Trichosanthes bracteata Voigt.* See p. 139, No. 43.

A few specimens somewhat behind the beach.

One should not forget that on this trip also a very small part only of the island was investigated. In the innumerable ravines which remained unexplored very many other plants may have been present. As the forest in the investigated parts was best developed at an altitude of 300 – 400 m. and from that altitude downwards diminished in the size of the trees, I think we are quite warranted to believe that its development has proceeded from the higher ravines downwards and that it was gradually invading the *Saccharum*-girdle from its inner side. Taking into consideration that the ravine-trees and the undergrowth belonged to such species as are frequent in the forests of Java and may quite well have grown on Krakatao before the eruption, further that the original soil in part of the higher ravines was laid bare very soon after the eruption, we have doubtless to reckon with the possibility (See p. 26) that this young ravine-forest, partly at least, was not introduced after the eruption but consisted of survivors of the catastrophe.

At the request of Dr. Treub I gave in the year-book of the Topographical Survey Service in the Dutch Indies 1908, pp. 189–191, a very succinct summary of the results of this trip. Having neglected to assiduously study Verbeek's most valuable book on Krakatao and not considering that Treub's investigation in 1886 had been carried out much too hastily and over a too small part than that its conclusions might be estimated to be valid for the entire island, I shared the opinion of everyone at that time that Krakatao had been buried in 1883 everywhere under a very thick and very hot layer of ashes and pumice. Starting from this erroneous premise I came to the conclusion that it was highly probable – though not fully proven – that the entire vegetation had been destroyed in 1883 and that the new vegetation had been wholly introduced. With the exception of this very badly founded conclusion I can still approve of the survey mentioned above. *Unhappily this conclusion is the cardinal point wherewith the Krakatao-problem stands or falls.*

On p. 26 seq. I have already given my present views on the origin of the present inland vegetation. It is not necessary to repeat here what has been said there.

I need scarcely say that from the trip of 1908 it does not appear, neither could appear, that the vegetation had grown richer since 1906. All plants found in 1908 for the first time may have been growing on Krakatao years before.

X. THE STAY OF Mr. JOHANN HÄNDL ON KRAKATAO FROM 1915 TILL 1917.

In October 1916 Mr. Johann Händl [1]) obtained from the Dutch Indian Government a concession for winning volcanic products on Krakatao. Having often visited the island after the eruption and having ascertained that it contained an inexhaustable store of pumice, Mr. Händl intended to use this material for the building of cheap and hygienic houses, for which he meant to have discovered a suitable process. The concession was granted for the eastern half of the island and covered an area of 870 ha. (the entire island is ± 1400 ha.). As Mr. Händl for reasons which it is not necessary to state here, was not able to effect his purpose and was in arrear with the payment of the yearly retribution due to the Dutch Indian Government, the concession was canceled in February 1921. In July 1919 the western half of the island had already been declared a Nature Monument to which in January 1925 the eastern half was added.

Partly from Mr. Händl himself and partly, through the medium of Dr. O. Arrhenius, from his son, I obtained the following information about Mr. Händl's prolonged stay on the island.

Mr. Händl with some relations and his native servants, in total 4 European families and about 30 coolies, established themselves on Krakatao towards the end of the year 1915; he himself stayed there till the end of the year 1917. During his stay the communication with Labuan (a small port on the western coast of Java) was kept up by means of a motor-boat, which went there frequently, usually once a week, in order to fetch the things necessary. On the south-eastern side of the island, in the *Casuarina*-forest, he built a house, the materials (wood and bamboo) for which were imported from Java. It would not have been necessary to import wood as plenty of dead trees were found washed ashore. So many trees were lying on the beach that the little community could collect abundance of firewood; nevertheless as a rule living trees were felled for fuel. Somewhat north of the house a kind of harbour was made. Next to his house Mr. Händl laid out a garden in which he cultivated i.a.

[1]) Mr. Händl was a German who had entered the service of the Dutch Indian Government in 1874 and had left it in 1895. He died on August 1st 1927 at Depok near Batavia.

coco-palms (*Cocos nucifera L.*) papaws (*Carica papaya L.*), cassava (*Manihot utilissima Pohl*), Guinea-pepper (*Capsicum annuum L.*), Chillies (*Capsicum frutescens L.*) and diverse species of vegetables and aromatic weeds (i a. *Gastrochilus panduratum Ridl.*). As domestic animals he kept hens, geese and swine. He sunk many shafts, one in order to get drinking water, the others in order to ascertain how thick the layer of pumice was in diverse localities. A well sunk quite near the house had a depth of six metres and yielded all the year round a sufficient amount of water. This water, though drinkable, had a peculiar taste, reminding of hydrogen sulphide (H_2S). When sinking a well quite near the coast, water was struck at a much less depth. Beneath the layer of ashes and pumice *Mr. Händl found in many places unburnt tree-trunks and other parts of plants.* He got convinced that the vegetation, partly at least, had not been burnt by the eruption of 1883 but only buried.

Off and on young coco-nuts were found washed ashore [1]) which, *when planted in a safe spot,* grew up into trees. — At an altitude of 600 or 700 meters tree-ferns were found.

During Mr. Händl's two years' stay on the island (1915—1917) *there never was a long period of drought.* The rains were but seldom extremely violent and never such as to cause serious trouble. Only during and shortly after heavy showers the ravines and gullies carried water to the sea. The top of the mountain was frequently wrapped in clouds.

By sailing vessels the island may be easily reached from June till August. During his stay on the island Mr. Händl received many visitors. After many fruitless efforts his son discovered a way to the top of the mountain, which could be reached only from the eastern side because there were not accumulated so many loose materials. He guided to the summit several of his visitors, i.a. Mr. Docters van Leeuwen and some officers of the Topographical Survey Service.

As to wild animals, he observed on the island many bats, pigeons, monitors, pythons, black and red ants and mosquitoes. Houserats and cockroaches were accidentally imported with luggage; the first rapidly multiplied. The sea around the island abounds in fish.

In 1917 Mr. Händl went back to Java but left some people on Krakatao in guard of his properties. From time to time he or his son revisited the island. His son was there when in April 1919 Mr. Docters van Leeuwen visited the island and also in

[1]) These may quite well have originated from the island itself.

October of the same year, when Krakatao was visited by the members of the First Dutch Indian Congress for Natural Science (See Chapter XI). After 1919 the house was deserted and fell gradually into decay, in October 1921 only a mass of ruins [1] was left.

On Oct. 6th 1919, after a rather long dry period [2], the island was visited by the members of the First Dutch Indian Congress for Natural Science. One of these members set fire to the dry grass in order to see whether it would burn. It *did* burn, to the great amusement of many present. Mr. Händl wrote to me about this case of arson: „Das Feuer griff so schnell um sich, „dass im Augenblick ein Teil der Insel einem Feuersee glich. Bei „dem Anblick dieses wild um sich greifenden Feuers, bei fürch- „terlichem Geknatter und Gekrach der brennenden Sträucher und „Bäume freute sich die Gesellschaft ausserordentlich, aber keiner „dieser Gesellschaft dachte, dass bei dieser trockenen Zeit in „einem Augenblick die ganze Insel in vollen Brand kommen könnte „und Menschen und Tiere mit meiner Unternehmung rettunglos leben- „dig verbrennen müssen. . . . . Mein Sohn, der uneingeladen von der „Unternehmung aus auf das Schiff ging, sah, dass man zurückfuhr

[1] Dammerman, The Fauna of Krakatao, Verlaten Island and Sebesy, in Treubia III (1923), p. 65.

[2] In 1919 the east-monsoon was in West-Java and South-Sumatra much drier than usually, as appears from the table herebeneath in which I have given the mean rain-fall and the mean number of rain-days in July, August and September for the rain-stations Anjer Kidool (West-Java) and Telok Betoeng and Kalianda, both in South Sumatra (See map).

		July	Aug.	Sept.	
Anjer Kidool (West-Java).	Mean fall (27 years):	88	88	61	mm.
	Fall in 1919:	3	57	8	mm.
	Mean number of rain-days:	5,8	6,3	5,3	
	Rain-days in 1919:	1	4	3	
Telok Betoeng (South-Sumatra).	Mean fall (44 years):	109	132	129	mm.
	Fall in 1919:	46	52	27	mm.
	Mean number of rain-days:	9,3	10	9,2	
	Rain-days in 1919:	8	8	9	
Kalianda (South-Sumatra).	Mean fall (10 years):	89	98	120	mm.
	Fall in 1919:	45	28	85	mm.
	Mean number of rain-days:	7,4	8,3	7,6	
	Rain-days in 1919:	4	4	3	

From 15 September till 5 October there fell in Anjer Kidool only 5 mm. (last rain-day 24 Sept.), in Telok Betoeng 10 mm. (last rain-day 26 Sept.), in Kalianda 17 mm. (last rain-day 24 Sept.). For Krakatao no definite figures are known but it is certain that there also was about this time a period of drought as appears from a letter [3] of Mr. Händl. He says: „Der Krakatau war gut disponiert für einen Waldbrand, die „Gräser, Sträucher und Bäume waren durch Mangel an Regen sehr trocken"

[3] Dated June 28th 1927.

„ohne die Menschen auf der Unternehmung von dem brennenden Eiland „abzuholen, befahl dem Schiffskapitän sofort zurückzukehren, um die „dortigen Menschen zu retten.

„Zwei Dinge kommen bei dieser Brandstiftung in Betracht. Einer- „seits wie es nur möglich ist, dass man ein so wertvolles Forschungs- „objekt der Neuentwickelung von Pflanzen und Tierwelt, das sicher „nicht leicht mehr vorkommt, so leichtsinnig der Vernichtung preis- „geben könnte. Anderseits möchte ich sagen, dass das eben genannte „wissenschaftliche Opfer zu gross und zu riskant, an Menschenleben „dabei noch nahezu unersetzlich war, um mich aus Krakatau herauszu- „brennen, denn es ist unmöglich anzunehmen, dass der gelehrte Brand- „stifter nicht eben so weit denken kann als der gänzlich ungebildete „inländische Fischer, der am Strande, weit von bewachsenem Grund, „die Brandgefahr fürchtend, sein Feuer zum trocknen seiner Fische „und zum Kochen macht, ganz abgesehen von strafbarer Handlung".

The bitterness with which Mr. Händl speaks of this impardonable imprudence by which his properties might have been seriously damaged, is quite justifiable. The vegetation on the island burned for days, the conflagration spread far up and around the mountain till finally the rains quenched the fire.

The long stay of Mr. Händl and his many followers on Krakatao, the frequent communication with Java, the import of many coolies and goods from there must – temporarily at least – have influenced the flora of the island as it has its fauna. A large clearing was made for a house and garden, paths were cut through the vegetation, many trees around the house were felled for fuel, domestic animals with part at least of their food, and plants or seeds for cultivation were introduced. Ruderal plants and weeds of agriculture, always following man and settling wherever he settles, must have come over [1]). The great fire of 1919 must have killed an immense number of herbs, shrubs and joung trees. The son of Mr. Händl informed Dr. Arrhenius that two years after the great fire its vestiges were still discernible everywhere.

Now that Krakatao has been definitively left by its inhabitants it is to be expected that the jungle will invade the old premises and kill out most of the cultivated plants and ruderal weeds. Perhaps this has already happened. If the island remains a Nature Monument, the

1) Though several such weeds were found in 1919 by Mr. Docters van Leeuwen and were recorded in Appendix I to his paper on the excursions of that year (See Chapter XI), he nevertheless does not make any mention in this paper of Krakatao being inhabited.

forest will gradually cover everything behind the beach, except localities unfit for it, as there are the basaltic rocks at Zwarte Hoek and other stony places and very steep walls. Only some coco-palms, planted and more or less protected by man, will in the future speak of human intervention. And probably the presence of these same trees will yet many a time be cited by persons, unacquainted with the history of the island and with the customs of the natives, as an irrefutable proof that Cocos *nucifera* can spread without human aid. History repeats itself. And nothing has a longer life than an error.

XI. THE EXCURSIONS OF DR. DOCTERS VAN LEEUWEN AND HIS COMPANIONS TO KRAKATAO IN 1919.

(Dr. W. Docters van Leeuwen, De flora en de fauna van de eilanden der Krakatau-groep in 1919 (met V bijlagen) in Handelingen van het Eerste Nederlandsch-Indisch Natuurwetenschappelijk Congres, gehouden te Batavia van 3–6 Oct. 1919 (1920), p. 35 seq. – The flora and the fauna of the islands of the Krakatao-group in 1919, in Ann. Jard. Bot. Buitenzorg XXXI (1921), p. 103 seq.).

From April 24th to 28th 1919 Messrs. Docters van Leeuwen (Director of the Botanical Gardens, Buitenzorg), Bartels (well-known ornithologist, head-manager of the tea-plantation Pasir Datar near Sukabumi (Java) and Sunier (Head of the Laboratory for Marine Investigations at Batavia) made an excursion to Krakatao. Mr. Docters van Leeuwen collected there a number of plants which some time after his return were named by officials of the Buitenzorg Herbarium. He stayed 3 days on the south-eastern side, made two trips into the interior and only one of these to the higher parts, which is altogether insufficient for a thorough exploration of the very dense vegetation of the numberless ravines and ridges. He did not investigate the slopes on the north-western side where Treub and Penzig had made their excursions.

At the First Dutch Indian Congress for Natural Science, held at Batavia in 1919, Mr. Docters van Leeuwen delivered on Oct. 4th a popular lecture on Krakatao and its flora and fauna. Two days afterwards the members of the congress paid under his botanical guidance the fatal visit to Krakatao spoken of in the preceding chapter. The lecture mentioned above, augmented with some appendixes, was in 1920 published in the Transactions of the Congress, p. 35 seq. In 1921 it was with some alterations published once more, now in an English translation in Ann. Jard. Bot Buitenzorg XXXI (1921), p. 103 seq. In these publications not any mention was made either of Krakatao having been inhabited by Mr. Händl himself during more than 2 years and afterwards by some of his followers, nor of the jungle-fire caused by a member of the congress. In the programma of the excursion to Krakatao [1]) it is even expressly stated that the

[1]) Handelingen Eerste Nederlandsch-Indisch Natuurwetenschappelijk Congres (1920), pp. (32), (33).

island was uninhabited [1]), though Mr. Docters van Leeuwen had been guided by the son of Mr. Händl (See the preceding chapter) who was then staying on the island. In a report [2]) published by Mr. Dammerman on the congress likewise a complete silence is preserved on the fire (of which fire Mr. Dammerman was an eye-witness) and on the fact that Krakatao had been inhabited [3]).

The lecture of Mr. Docters van Leeuwen having been inserted by its author in extenso in a scientific paper and containing the newest data published on the vegetation of Krakatao, has to be discussed here. As it contains, besides these data, rather many inaccuracies and a number of views and speculations that cannot possibly be agreed with, a somewhat detailed account of it is necessary. Therefore I give herebeneath a synopsis of it in which I adhere to the English text as closely as possible. Wherever this text notably differs from the original Dutch one I shall refer to the latter also. At the end of each paragraph I give the necessary comments.

§ 1. By the eruption of 1883 not only all that lived on Krakatao was absolutely annihilated but, in addition, the entire surface was shrouded with a thick layer of burning-hot ashes, so that it may be said that the soil was sterilized. On considering the present state of the flora, when one sees the gigantic *Ficus*-trees, one is involuntarily inclined to wonder whether everything was really destroyed or whether some individuals might not have escaped destruction. We have however the statements of Verbeek and Treub, who were, so to say, on the spot and we are bound to accept their statements as to the absolute extinction of the entire vegetable life, unless facts of a serious nature should be discovered which would justify doubts as to the reliability of them. Such facts have not been found.

This subject having been already sufficiently discussed in Chapter III, I refer to what I have said there on the conclusions made on this point by Verbeek and Treub. One should not forget that Treub's first visit took place in 1886, 3 years after the eruption,

---

1) In a subsequent paper (W. Docters van Leeuwen, The Vegetation of the Island of Sebesy, situated in the Sunda-Strait near the Islands of the Krakatau-group; in the year 1921, Ann. Jard. Bot. Buitenzorg XXXII (1923), p. 142) Mr. Docters van Leeuwen gives some information about Krakatao having been inhabited but this information is far from accurate. He says: „since 1917 a European settled there with „a number of coolies on the N. E. side. The original vegetation was not however „interfered with to any extent, a small piece of forest only being felled in the „immediate vicinity of the dwelling-house and a few plantations started. But from 1919 „onward the house has been deserted again and everything has been grown over „once more". — As a matter of fact Mr. Händl settled on Krakatao not in 1917 but in 1915, not on the N. E. but on the S. E. side, and the vegetation was certainly much interfered with, many trees having been felled and most probably several ruderal weeds having been introduced. In this paper also Mr. Docters van Leeuwen keeps silent on the great fire.

2) Vakblad voor Biologen I (1920), p. 71 seq.

3) With this fact Mr. Dammerman was quite well acquainted as appears from his paper in Treubia III (1923), p. 65.

so that he can hardly be said to have been *on the spot* at the time of the eruption. The *Dutch* text cites as yet another witness of the total destruction in 1883 the zoologist Sluiter. In 1888 and 1889, 5 and 6 years after the eruption, this savant investigated the coral reefs [1]) of Krakatao but had, as he himself stated [2]), no time to investigate the new terrestrial fauna. It does not appear that he ever has ascended the mountain or instituted any investigations after the new flora, nor has he, to my knowledge, published anything about it. Therefore it is inconceivable that his testimony has been invoked. — For newly discovered facts of a serious nature I refer to the preceding chapter, p. 197, where mention is made of the discovery of *unburnt* treetrunks and other parts of plants beneath the layer of ashes and pumice.

§ 2. The loose materials accumulated by the eruption has *already* [3]) been washed away here and there and especially in the ravines and gorges the original rock has come to the surface again and in those parts one can indeed find charred and rotted tree-boles, but everything else came into existence after 1883.

This communication does not possess the full charm of novelty, its essential points having already been published 35 years before by Verbeek [4]), who two months after the eruption found the original soil to lie bare in the eroded ravines and who also mentions the presence of fallen and carbonized treetrunks. As to the statement that everything else (by which, apparently, only the vegetation and the fauna are meant) came into existence after 1883, I refer to Chapter III, herebefore.

§ 3. The development of the new flora has not received all the attention it deserved at the hands of naturalists. Treub visited the island in 1886. We lack information as to the first beginnings of the new flora but still the results of Treub's visit are very important. In 1897 Treub, accompanied by Penzig and some other naturalists, came to Krakatao a second time. The flora was then already comparatively rich. About what happened in the interval of years between those two visits one cannot do more than form a general notion. In 1906 the island was visited by Ernst and his companions, in 1908 by an expedition of the Topographical Survey Service, who had been joined by the first zoologist Jacobson and the botanist Backer. In 1919 Docters van Leeuwen came together with the zoologists Bartels and Sunier. It was only on the last of all these expeditions that the summit was reached by a *scientific* [3]) investigator so that our knowledge of the part of the

---

[1]) C. Ph. Sluiter, Einiges über die Entstehung der Korallenriffe in der Javasee und über neue Korallenbildung bei Krakatau [Nat. Tijdschr. Ned. Indië XLIX (1890) p. 360 seq.].

[2]) Nat. Tijdschr. Ned. Indië XLVIII (1889) p. 353.

[3]) Italics by me. B.

[4]) Verbeek, Kort Verslag (1884) p. 13; Krakatao (1888), pp. 122, 475,

mountains upwards of 1200 ft. high is of very recent origin, for Mr. Demmeni, the photographer of the surveying-party, who reached the top in 1908, could give but scanty information concerning the flora.

This statement is correct except as to the height reached in 1908, which the Dutch text in the Transactions of the Congress rightly gives as 400 m. or about 1350 English feet.

§ 4. On April 24th 1919 Mr. Docters van Leeuwen landed on the south-eastern side. He expected to see plant-growth but that it should be so full was still a surprise. The first impression is of a vegetation similar to what may be found on other islands [1]): a fringe of coco-palms and higher up the dense forest. On Krakatao the border of coco-palms is replaced by a belt of *Casuarina equisetifolia L.* There are indeed coco-palms but they are not yet very prevalent. The *Casuarinas* form a broad zone, the trees being already quite tall and robust and entirely overgrown by *Ipomoea denticulata Chois.*, *I. longiflora R. Br.* (= I. grandiflora Lamk.) and *Columella trifolia Merr.* (— Vitis trifolia L.). The shore is grown with the usual littoral plants, *Ipomoea pes caprae Sw.*, *Canavalia rosea D. C.* [2]) and *Vigna marina Merr.*, which together with *Ischaemum muticum L.*, *Spinifex littoreus Merr.*, and other herbs form the *Pes-caprae*-formation. Everywhere teems *Cassytha filiformis L.* Immediately behind the white beach there is a low wall of shore-cliff grown with various plants which Mr. Docters van Leeuwen thinks he need not enumerate all and immediately behind this follows the *Casuarina*-forest. The belt of *Casuarina*-forest consists of various kinds of trees, belonging to the *Barringtonia*-formation, which owes its name to *Barringtonia asiatica Kurz.* To the same formation belong *Terminalia catappa L.* and *Hibiscus tiliaceus L.* Birds were heard singing; the oriole was audible everywhere. The *Casuarina*-forest has a thick undergrowth of various bushes and plenty of trailing and creeping plants, so that it is almost impenetrable. Behind the *Casuarinas* there is a broad stretch consisting of various grasses among which *Saccharum spontaneum* is most in evidence. — On his walk along the beach Mr. Docters van Leeuwen penetrated here and there into the wilds and collected about 100 different plants, which were carried home, labelled and dried.

Mr. Docters van Leeuwen was astonished to see on Krakatao such a dense vegetation. Whoever takes the trouble to read what Ernst [3]) as early as 1907 published on this vegetation (See Chapter VIII, p. 143) will be astonished at this astonishment. The statement of the aspect of the vegetation in 1919 is neither clear nor accurate. It seems that Mr. Docters van Leeuwen meant to say that on Krakatao behind the flood-mark lay the beach covered with a *Pes caprae*-formation, behind this the *Barringtonia*-formation and behind the latter the *Casuarina*-forest. We are not told whether in 1919 this *Casuarina*-forest was present on the entire eastern side of the island; in 1906 and 1908 *Casuarina*

[1]) Here the Dutch text with a fine sense of the picturesque adds: „and which was characterised by a jovial sea-captain by these words: *„first a white beach, then coco-palms, for the rest borecole".*

[2]) Often confused with C. *lineata D.C.*

[3]) Neue Flora Vulkaninsel Krakatau (1907), p. 28; p. 32 seq.

occurred chiefly at the south-eastern angle; farther to the north only scattered specimens were present; the grass-jungle lay there either immediately behind the beach or, wherever the *Barringtonia*-formation had developed, behind the latter. It is however quite possible that the *Casuarina*-forest has since spread or shifted to the north.

In § 15 hereafter Mr. Docters van Leeuwen states that here and there the grassy plains advanced as far as the littoral zone. Whether from this statement is to be concluded that in these places the *Casuarinas* were lacking is not clear; sometimes it seems that *Casuarina equisetifolia*, which is considered by [1]) Schimper as a component of the *Barringtonia*-formation, but which had better be regarded as making a formation of its own — is by Mr. Docters van Leeuwen not even considered as a littoral plant (Cf. § 15). I cannot agree with the author where he says that on Krakatao *Casuarina equisetifolia* replaces the fringe of coco-palms „on other „islands". Wherever coco-palms are found on those other islands they have originally been planted and, as a rule, not in „a fringe" (though it may seem so when seen from afar) but in several rows and at some distance from the sea. Only where the beach gets washed off by the waves, coco-palms may be found immediately along the border of the sea. — *Casuarinas* on the contrary are never cultivated on coral-islands [2]), but grow spontaneously on well-pervious beaches rich in lime such as are regularly found on such islands.

In 1906 the highest *Casuarina* on the south-eastern side of Krakatao measured about 25 m. Mr. Docters van Leeuwen calls the trees in 1919 quite tall and robust but gives no figures. But in a paper [3]) published one year afterwards he says that the tallest trees were more than 33 m. high. I myself found *Casuarinas* of about that size already in 1908 [4]). As plants entirely overgrowing the tall *Casuarinas* are mentioned *Ipomoea denticulata Chois.*, — *I. longiflora R. Br.* and *Columella trifolia Merr.* This cannot possibly be correct. — *Ipomoea denticulata* is a comparatively slender twiner, which never reaches a great height; it cannot have entirely overgrown tall *Casu-*

---

[1]) Schimper, Indo-Malayische Strandflora (1891) p. 69.

[2]) Only on the calcareous island of Madura I saw on the sand-dunes of Batang batang daja a wood of *Casuarina equisetifolia*, planted by the natives. The trees were all polled quite beneath; the young shoots were regularly cut off for fuel.

[3]) W. Docters van Leeuwen. The galls of Krakatau and Verlaten Eiland in Ann. Jard. Bot. Buitenzorg XXXI (1921), p. 62.

[4]) See herebefore, p. 187, and Jaarverslag van den Topographischen Dienst in Nederlandsch Indië (Year-book of the Topographical Survey Service in the Dutch Indies) 1908, p. 190.

arinas. — *Ipomoea longiflora* is much more robust but I never saw it at a height of at least 20 meters form the dense carpet requisite for entirely covering a tall tree. *Columella trifolia,* on the contrary, as Ernst[1]) already rightly mentioned, can reach a great height and form there a dense canopy.

It is much to be regretted that Mr. Docters van Leeuwen does not mention a single one of the „various bushes and plenty of „trailing and creeping plants" which formed a thick undergrowth of the *Casuarina*-forest. — Of the „usual plants" of which the *Pes caprae*-formation was composed he mentions only six; it is not clear why the other ones need not be enumerated in a paper intending to sketch the development of the flora. It would surely be of interest to know whether the *Pes caprae*-formation on this side of the island had grown much richer in species since the visit of Ernst. Appendix I, giving all the plants found in 1919 on Krakatao, cannot help us here. It is remarkable that whilst in § 4 *Canavalia* is given as one of the plants with which the shore was overgrown, this species is not mentioned for this trip in Appendix I, giving all the plants collected on Krakatao. Of the components of the *Barringtonia*-formation but 3 are recorded and these only in a general way; many typical and conspicuous trees and shrubs of this formation [2]) which Ernst and his companions had found on this side of Krakatao in 1906 (see the list on p. 153 seq.) are not mentioned. It is true that in Appendix I to Mr. Docters van Leeuwen's paper most of these species are recorded as found on Krakatao in 1919 but this appendix gives no information on which side of the island and in what numbers they occurred, so that it is impossible to compare the littoral vegetation of 1906 with that of 1919.

The belt of grasses lying behind the *Casuarinas* (in 1919 never immediately behind the beach or behind the Barringtonia-association? see § 15) consisted of various species of which *Saccharum spontaneum L.* was „most in evidence". We are not told whether this belt was everywhere dense, neither which other plants than *Saccharum* occurred in it. It is apparently deemed of more importance for science to tell that about 100 species were found, labelled and dried, than to mention which these species were and in what habitat and in which numbers they occurred. *It should never be*

[1]) Ernst, Neue Flora Vulkaninsel Krakatau (1907), p. 30.

[2]) i.a. *Calophyllum inophyllum L.,* — *Desmodium umbellatum D. C.,* — *Dodonaea viscosa Jacq.,* — *Erythrina variegata L.* var. *orientalis Merr.,* — *Guettarda speciosa L.,* — *Hernandia peltata Meissn.,* — *Morinda citrifolia L.,* — *Pandanus tectorius Sol.,* — *Pongamia pinnata Merr.,* — *Scaevola frutescens Krause,* — *Sophora tomentosa L.*.

forgotten that any person in a favorable locality in the tropics can collect a great number of species within a few hours. But in the case of Krakatao such a collection, not accompanied by reliable oecological data, is of exceedingly little value.

§ 5. On April 25th an exploration-trip was made in the lower ravines for the purpose of seeking a track to the top. First the way follows a water-course in the field of *Saccharum spontaneum L.,* swarming with thousands of ants. After that a lovely spot was reached, the gully of a dry water-channel, with many tall trees and a good deal of underwood and fern. Many trees there were belonging to a few species, especially the *Ficus*-family, i.a.: *1. Ficus fistulosa Reinw.,* – *2. F. fulva Reinw.,* – *3. F. variegata Bl.* – some of them quite big; furthermore *4. Radermachera glandulosa Miq.,* – *5. Homalanthus populnea O. K.,* – *6. Macaranga tanarius M. A.,* – *7. Nauclea purpurascens Korth.,* – *8. Melochia umbellata Stapf,* – *9. Trema orientale Bl.* – *10. Pipturus incanus Wedd.,* – *11. Villebrunea rubescens Bl.,* some of the specimens so robust that one would certainly have assigned an older age than 36 years to them, if one had not known the history of the island. It was a peculiar kind of virgin forest such as is rare in Java, since it was made up by many individuals representing but few species. Epiphytic ferns grew everywhere, a.o. good-sized *Asplenium nidus L.* This also was remarkable, because Backer mentions that epiphytes were still rare in 1908. The great variety of mushrooms and toadstools vegetating on the fallen treeboles was very striking, the more so as Ernst only mentions 2 of these. Mr. Docters van Leeuwen also found several new species of ground-orchids, a. o. *Habenaria* spec., *Eulophia macrostachya Lindl., Nervilia aragoana Gaud.* The species most abundant at the time of former explorations, *Arundina speciosa Bl.* and *Spathoglottis plicata Bl.* now proved to have been ousted by the spreading forest. The zoologist collected a number of lob- and lugworms (*Lumbricus* spec.) which had not been perceived in 1908. – At an altitude of about 100 m. the way was blocked by an absolutely unscalable precipice. About 50 plants were collected at all.

Mr. Docters van Leeuwen does not mention from which point he tried to penetrate into the interior. In 1908 the investigators (see Chapter VIII) started from the south-eastern angle. It is next to impossible that Docters van Leeuwen in 1919 followed quite the same route and examined the very same ravines as his predecessors had done. The track made in 1908 must within a few years have been entirely shut up by the dense tropical jungle and could not possibly be rediscovered. Referring to what I have said on p. 5–7 about the composition of comparatively young vegetations in the tropics I need scarcely argue here that from a comparison of the finds in 1906 or 1908 with those in 1919 no reliable conclusion can be drawn about the appearance or disappearance of individual species, though we are fully warranted to believe that – generally speaking – the young forest had extended still more and probably considerably downwards.

From the narrative of Dr. Docters van Leeuwen, who first followed a water-course and afterwards a dry water-channel one would be inclined to deduce that he found living water on the island. But as such water — except during and shortly after heavy rains — was never seen by Mr. Händl, who was very well acquainted with this side of the island (See Chapter X), and as in the next paragraph it is stated that water was nowhere to be had, I rather think that also the first water-course was waterless. This is a point of importance: the flora along permanent water-courses always differing from the rest.

Dr. Docters van Leeuwen is quite wrong where he says that a forest as found by him on the lovely spot in the dry water-channel and consisting of but few species, is rare in Java. On the contrary such forests are frequent in thinly populated regions where, in order to obtain temporary rice-fields, part of a forest is cut down by the natives, who, after the fallen trees have been burnt and the otherwise not manured field has yielded a few crops, desert the spot and destroy another part of the forest for the same purpose. In the southern half of the Preanger Regencies and of Bantam (Western Java) this often happens. The secondary forests, developing on such deserted fields in South-Bantam were already in 1913 shortly described by me about as follows (translated from the Dutch [1]) original with some slight alterations):

„On a travel which Mr. Clerx and I made a few years ago „(1911) in the south-eastern part of the Residency of Bantam, the „appearance and disappearance of *Lantana Camara L.* could very „well be observed. When the native of South-Bantam leaves his „exhausted hooma (dry rice-field) to its fate, first tens of thousands „of fruits of alang-alang (*Imperata cylindrica P. B.*) arrive soaring on „the wind and soon turn the spot in a rather monotonous grass-field „in which however many other plants, for the greater part incon„spicuous, are growing. Numerically these latter form at first but „a slight minority. But after a short time shrubs begin to spring up „between the grass, especially shrubs with fleshy or juicy fruits, „which points to introduction by animals. *Melastoma malabathricum L.*, — „*Clerodendron serratum Spreng.* and especially *Lantana Camara L.*, „appear everywhere. The last species finally often wins, for it grows „higher and bears much more fruit than the other ones. Everywhere „one sees birds pecking its fruits; everywhere its seeds may be „found in bird-droppings. Moreover it does not suffer from leaf-

[1]) De Tropische Natuur II (1913), p. 30.

„eating Coleoptera which often sorely damage the two other spe-„cies, especially *Melastoma*. And soon everywhere dense *Lantana*-„thickets are seen, which may grow to a height of $1^1/_2 - 3$ m. and „finally kill the alang-alang by their shade.

„But the victors do not remain a long time in the undisturbed „possession of the territory conquered. For between the shrubs trees „begin to grow up, especially species of *Vitex, Grewia, Dillenia* „and *Hibiscus*. In the shade of these trees *Lantana* cannot thrive; it „languishes and finally disappears; the trees get the upper-hand and „finish by forming a dense forest [1]). In this way the monotonous „young secondary forests of South-Bantam originate, sunny and rather „dry, poor in species, lacking in humus, poor in undergrowth, poor „in epiphytes. For hours one may walk in them with always the same „plants, the same birds along the path".

This description applies to *young* secondary forests. As the forest grows older and higher it gradually becomes less dense, the smaller trees being partly killed out by the larger ones. The soil grows richer in humus and in moisture; gradually there develops an undergrowth of other plants; epiphytes settle on the trees, especially on the larger ones. The older the secondary forest is, the richer — ceteris paribus — it is in species. But many years have to elapse before the epiphytical flora can vie in richness with that of a primeval forest under the same climatological circumstances. Especially *Hymenophyllaceae*, which constantly require much moisture and much shade, are often slow in coming.

It is to be regretted that Mr. Docters van Leeuwen has neglected to state of which species the „undergrowth and fern" of the secondary forest examined by him, was composed. Of the 11 components of the secondary forest mentioned, 7 had already been found on this side of the island in or before 1908. No. 9, *Trema orientale Bl.* had in 1905, 1906 and 1908 been collected on the other side of Krakatao, near Zwarte Hoek. These species were already discussed in the preceding chapters.

No. 3, *Ficus variegata Bl.*, the third species belonging to the Ficus-family (recte: *Moraceae*), is a tree which may reach large dimensions (often 20—40 m. high; $1 - 1^3/_4$ m. thick); it bears its figs in clusters on the smooth trunk. It is rather common throughout

[1]) But this only happens where the young forest does not suffer too much from its three most dangerous enemies: wood-cutting natives, grazing cattle and fire. Where cattle daily grazes, the secondary growth is frequently almost restricted, besides to grasses, to *Breynia, Brucea, Harrisonia, Lantana, Macaranga, Melastoma* and a few other shrubs and herbs.

Java from the plains up to an altitude of ± 1500 m.; it thrives as well in regions where the east-monsoon is very dry as where it is very feeble. It grows in primeval and in secondary forests and in low jungles, often in ravines and on water-sides, but never in saline soils. The rather large figs are greedily eaten by monkeys, bats and birds, who most probably contribute to the dispersal of the very small drupes.

No. 4, *Radermachera glandulosa Miq.*, is much less wide-spread. It occurs from West- to East-Java, from the plains up to 700 m. but only in rather fertile and humid soils, especially in dense forests, generally in scattered specimens. The reproduction is effected by the long-winged seeds.

No. 11, *Villebrunea rubescens Bl.*, is a shrub or small tree, 3—8 m. high, spread throughout Java from the plains up to an altitude of 1600 m. but only in regions where the east-monsoon is rather feeble. It grows by preference in well-shaded localities, in primeval and in secondary forests, often in ravines. It reproduces itself by its juicy white fruits.

All these species may have grown in the former woods of Krakatao. Whether this has been really the case, whether they have survived the eruption or have been introduced after it, no one can tell.

Of the epiphytes growing everywhere only a single one is mentioned, *Asplenium nidus L.*, a well-known „nest-fern", common throughout Java in well-shadowed localities from the plains up to ± 2000 m. above sea-level, especially above 250 m., but only in regions where the east-monsoon is not very strong. Its frequent occurrence in the lower part of Krakatao proves that there, as a rule, rather much rain falls in the east-monsoon. The silence preserved about the other epiphytes is quite out of place in a study on the development of the new flora. It would be of interest to compare the epiphytical vegetation of the lower parts of Krakatao with that of the higher parts. Now this is impossible. The number of species found on this trip is stated, but the species themselves are not recorded and oecological data are almost entirely lacking. In the lower regions of Krakatao epiphytes were in 1919 much more common than in 1908 or more scientifically spoken, epiphytes were more common in an old secondary forest than in a young one. This is in Java a fact of common occurrence and needs not to be wondered at.

As to fungi, E r n s t found his 2 specimens on a single dead tree lying on the sunny beach. Mr. D o c t e r s v a n L e e u w e n, contrariwise, collected his fungi on fallen tree-trunks in a rather moist and dense forest, a locality much better fit for the development of such organisms. He mentions in his paper *mushrooms* and *toadstools* probably meaning by the first of these popular terms edible species, by the second inedible ones. Such a mere culinary distinction has to be considered out of place in a phyto-geographical study. As to the great variety of these mushrooms, the list of fungi given in the first Appendix includes but 3 species known to be edible: *Schizophyllum alneum Schröter* (= Sch. commune Fries); *Auricularia auricula judae Schröter* and *Polyporus grammocephalus Berkeley*, whilst a fourth species, *Pycnosporus sanguineus Murrill* (= Polystictus sanguineus Fries) is a native medicine. But of these 4, only *Schizophyllum alneum*, being somewhat fleshy, might with some show of right be named a mushroom, so that there is no question of a „great variety" of these plants.

Of the 3 new species of terrestrial orchids which were found in the dry water-channel, one [1]) *(Nervilia aragoana Gaud.)* had already been collected on the same side of the island in 1908; it is recorded for this trip by Mr. Docters van Leeuwen himself in Appendix I to his paper. *Arundina speciosa Bl.* and *Spathoglottis plicata Bl.* being heliophilous kremnophytes (See p. 107), one need not wonder that these species were not found in the dense forest. But it is by no means proven that they were not present elsewhere; the less so because Appendix I mentions them both as found in 1919.

The most remarkable discovery of „lob- and lugworms" on Krakatao appears to be due only to an error in the translation of the Dutch into the English text. For the lob- or lugworm, *Arenicola marina Malmgr.* (= A. piscatorum Lamk.) recognizable i. a. by a row of tufted gills along each side of the back, is a chaetopode only found burrowing in sandy beaches between the tide-marks [2]). But as may be derived from the addition *Lumbricus* sp. and from the Dutch text, no *lob*-worms were found by the zoologist but *earth*-worms. From their presence may be inferred that the soil in the forest was rather moist and not lacking in humus.

---

[1]) Mr. Docters van Leeuwen [Handelingen Eerste Ned. Ind. Natuurw. Congres (1920), p. 62, No. 141 and Ann. Jard. Bot. Buitenzorg XXXI (1921), p. 131, No. 141] mentions *Eulophia macrostachya Lindl.* (see herebefore p. 207) also for the expedition of 1908. This is an error, the species collected in that year was *Eu. macrorrhiza Bl.* See p. 180.

[2]) Hence the Dutch name *zeepier* (sea-worm).

§ 6. On April 26th Dr. Docters van Leeuwen attempted once more to ascend the top, this time in the company of Mr. Händl Jr., a native plant-collector and two coolies. The ascent was difficult, not only because of the steep acclivities and the loose quality of the soil, but also because of the moist heat prevailing in the densely grown ravines. Water was nowhere to be had. Such steep acclivities, sometimes amounting to more than 30%, are usually found only in the higher mountains where the cooler climate renders the ascent less exhausting; on Krakatao the steeps began at slight elevations above the sea. Everytime the end of a ravine was reached, whereupon the investigators had to scramble up obliquely against the crumbling wall to strike another gorge reaching a little higher upwards. At eleven o'clock they were still in the forest but a change in its composition was already perceptible. The same trees still occurred as those beneath, but some species became more prevalent, like *Villebrunea rubescens Bl.* and *Ficus ribes Reinw.*, which also in Java belong more properly to the mountainous regions. The most remarkable thing was that, as the explorers rose higher, the forest was almost wholly constituted of *Cyrtandra sulcata Bl.* The plants stood side by side in thousands, each consisting of a number of strong branches proceeding from a common centre and reaching a height of 3–4 m. At a height of ± 600 m. between the *Cyrtandras*, on which a host of epiphytes had fastened, typical mountain-trees were growing, i.a. *Ficus ribes Reinw.* The whole forest began to wear a moister aspect; from the branches depended heavy festoons of a kind of scale-moss [1]), that were full of water. At last the explorers reached a point where a huge specimen of *Ficus retusa L.* [2]) rose high above the monotonous vegetation (chiefly composed of *Cyrtandra sulcata Bl.*), that seemed to cover all the upper half of the mountain. With its thousands of aerial roots [3]) this tree covered a large area. The humidity increased, the ground got wetter and more slippery, but between the ravines were dry, often narrow ridges, sometimes interrupted by cross-grooves, and consisting of granular matter sliding away underfoot and held together by *Saccharum spontaneum L.*, *Imperata cylindrica P. B.* (= I. arundinacea Cyr.), but especially by the ferns *Nephrolepis exaltata Schott* (= N. hirsutula Presl) and *Dryopteris unita O. K.* (= Dr. cucullata Christ.). At half past twelve the verge of the rupture just below the top was reached. The trees began to look stunted and in some places there were already patches of *Saccharum spontaneum* but *Cyrtandra sulcata* still dominated, rather smaller indeed but otherwise apparently very little the worse for the fierce light. At half past one the explorers reached the top, which was entirely covered with *Saccharum spontaneum L.* and with 2 very tall species of ferns, *Nephrolepis exaltata Schott* and *Dryopteris unita O. K.* over which crept *Trichosporum volubile Nees. Cyrtandra sulcata Bl.* was still present, likewise *Ficus ribes Reinw.* and *F. fistulosa Reinw.* The grass reached above-head – The whole ascent had

---

[1]) The English text has *seale-moss*, which apparently is a misprint for scale-moss, i.a. a representative of the order *Hepaticae* (Dutch: *levermos*). The Dutch text however has *loofmos*, i.e. a representative of the order *Musci Frondosi*. Probably the reading of the Dutch text is right: at least Max Fleischer [Beitrag zur Laubmoosflora der Vulkaninsel Krakatau in Ann. Jard. Bot. Buitenzorg, XXXIII (1923), p. 107] mentions as a species found on this trip by Mr. Docters van Leeuwen *Floribundaria floribunda Fleisch.* as „in feuchten Schluchten, von den Aesten und Stämmen herab- „hängend 26/4/'19. Bildet anscheinend Massen-vegetation wie in den feuchten Berg- „wäldern der Javanischen Vulkangebirge". – Everyone who has visited the West-Javan rain-forests has remarked the dense festoons of *Aërobryum, Aërobryopsis* or *Floribundaria*, pending from the trees. See also Junghuhn, Java, 2nd Dutch Edition I, 525; German Edition I, 382.

[2]) As appears from Appendix III to the Dutch text this tree was growing at an altitude of ± 700 m.

[3]) The English text has here aerial *shoots*. From a comparison with the Dutch text it appears that aerial *roots* are meant.

required more than six hours. At two o'clock the descent began, which was finished by four o'clock. No fewer than 50 species of plants had been found, nearly all of them undiscovered in previous explorations. Sunier had collected (where ?) many animals, Mr. Bartels had found the birds to be extremely timid.

Mr. Händl Jr. who during his long stay on the island had succeeded, after several fruitless attempts, in finding a way to the top and who that day guided the party, informed me through the medium of Dr. O. Arrhenius, that the ascent took place from the north-eastern angle, which can be very easily reached from the south-eastern point by following the beach. The steep acclivities the explorers had to ascend shortly after having set out on their trip were not the gentle slopes of the lower part of the vulcanic mantle itself but steep parts of the bottoms or the sides of the eroded ravines. Such ravines are in Java very frequently found quite near the coast, but may be not recognized as such, because, as a rule, a river flows through them and often nearly fills them up. Such is not the case on Krakatao. In basing conclusions on the botanical results of this trip one should bear in mind that the vegetation of only a very small section of the higher parts of the mountain was examined, viz. the margins of the path followed, and that it is quite possible that in ravines on the other side of the mountain the vegetation differed very much. Furthermore, that a period of six hours was much too short for a thorough investigation of the dense vegetation along the way; very many plants must have been overlooked; the data acquired cannot but be very incomplete. Yet a great many valuable observations might have been made by an able botanist. It is much to be regretted that Mr. Docters van Leeuwen gives so very few details on the composition of the vegetation seen by him. He mentions few plants and on these he procures but very scanty data, often not even the vertical range of distribution. In Appendix III to his paper one finds, it is true, a list of forest-plants (afterwards to be discussed) and to most of the names contained in that list the altitude is added, but the height given apparently often bears only on the spot where a *specimen* was collected. The list would be much more useful if it contained the vertical range of each of the *species* found. Now one is not even able to form an idea of the distribution of the plants along the path and these form but an infinitesimal fraction of the whole. Of the distribution of species over the rest of the mountain we know nothing at all.

At 11 o'clock a point was reached where a change in the composition of the forest became perceptible; *Ficus ribes* i.a. grew more frequent. As the height of this point is not given and nothing is told about the composition of the lower part of the forest examined that day, this statement does not make us much the wiser. But perhaps it is implied that the composition of this forest was about the same as of that examined the previous day (See § 5, p. 207), for which however *Ficus ribes* is not recorded. The statement that in Java *Villebrunea rubescens* more properly *belongs* to the mountainous zone is not quite correct; it is apparently based on confusion of human influences with natural ones. This species is a rather common forest-plant in regions *where the east-monsoon is feeble;* there it is found from the plains up to rather high in thc mountains. *But in the lower parts of Java such forests, as a rule, have been destroyed by man.* Hence it *seems* to the insufficiently trained explorer that this plant belongs, i.e. *is by Nature restricted* to the mountainous zone. But where primeval or old secondary rain-forests are present in constantly moist regions *Villebrunea* is often frequent at low altitudes. *Ficus ribes*, a small tree, requiring like *Villebrunea* a feeble east-monsoon is spread throughout Java but especially in the western and central parts between 100 and 1600 m. in moist forests. It is certainly much more common in the mountains than in the lower regions, but yet it has at several places been found at altitudes beneath 600 m. Koorders and Valeton[1]) mentioned this species already a quarter of a century ago for altitudes of 100 m. Therefore it may hardly be called a *typical* mountain-tree. This denomination I should wish to reserve for such species as are *exclusively* found on the mountains above an altitude of 600 m.

It is certainly remarkable that *Cyrtandra sulcata* was so very common. It would be of interest to investigate which was the cause of this phenomenon. This seems not to have been done. In Appendix I to his paper Mr. Docters van Leeuwen affirms that the plant is spread by animals, but it does not appear that this assertion is based on actual observation; it is probably only deduced from the fact that *Cyrtandra* has fleshy fruits. Even if animals were observed to eat the berries, this would not yet furnish proof that *only they* are the agents of dispersal of the minute seeds. Ripe berries frequently fall off and rot away; rainwater may spread the free-coming seeds. In a

---

[1]) Bijdrage XI tot de Kennis der Boomsoorten van Java (Contribution XI to the Knowledge of the Javanese Trees) 1906, p. 203.

subsequent publication[1]) Mr. Docters van Leeuwen says that *Cyrtandra sulcata* on Krakatao „*completely and densely covers all slopes „between 300 and 800 m.*". But it can hardly be believed that during his few and short visits to Krakatao he has found time to visit all or nearly all the slopes, many parts of which are inaccessible by the presence of innumerable deep and vertical-sided ravines, intercrossing in all directions. Neither is it probable that *ridges, ravine-sides* and *ravine-bottoms* alike were *completely* and *densely* covered with one and the same species.

Of the host of epiphytes which had settled on the *Cyrtandra* at a height of ± 600 m. *not a single species* is mentioned. But in the list of forest-plants added as Appendix III to his paper, Mr. Docters van Leeuwen records 9 epiphytes (*Liparis viridiflora Lindl.,*— *Oberonia monstruosa Lindl.,*— *Oberonia* spec.,— *Thrixspermum* spec.,— *Trichomanes humile Forst.* [2]),— *Antrophyum* spec.,— *Davallia* spec.,— *Hymenolepis spicata Presl* and *Polypodium heracleum Knze*) for a height of 600 m. or more. The bryologist Max Fleischer[3]) mentions the moss *Floribundaria floribunda* Fleisch., as having been found on trees between 200 and 700 m. on the day of this excursion. For a tropical rain-forest at that altitude a total of 10 species of epiphytes is but a very small number but many more may have been present. And it should not be forgotten that many epiphytes are slow in coming to a secondary forest.

Of the typical mountain-trees occurring at an altitude of 600 m. only one, *Ficus ribes Reinw.*, is mentioned and this is, as I already observed herebefore, hardly a *typical* mountain tree, because it is also found in the lower regions. Whether the huge specimen of *Ficus retusa L.* found at an altitude of 700 m. (as appears from Appendix III to the Dutch text) was a remnant of the former vegetation or had grown up after the eruption, cannot be ascertained any more.

On the drier ridges heliophilous plants occurred, part of them common kremnophytes (*Nephrolepis exaltata Schott, Dryopteris unita O.K.*),for the rest two very common grasses. The frequent occurrrence of *Saccharum spontaneum L.* on the ridges, the multitude of epiphytical

---

[1]) W. Do ters van Leeuwen, The Vegetation of the Island of Sebesy, situated in the Sunda-Strait near the Islands of the Krakatau group; in the year 1921, in Ann. Jard. Bot. Buitenzorg XXXII (1923), p. 139.

[2]) Often this is no epiphyte but an inhabitant of wet rocks and steep walls. See the list of forest-plants hereafter, sub No. 1.

[3]) Beitrag zur Laubmoosflora der Vulkaninsel Krakatau in Ann. Jard. Bot. Buitenzorg XXXIII (1923), p. 107. The same paper mentions *Aerobryopsis longissima Fleisch.* (inclusive of var. *densifolia Fleisch.*) as having been found on forest trees at altitudes of 500—700 m. on Jan. 1st 1922 and. Febr. 5th 1922.

plants, the occurrence of the hygrophilous *Trichosporum volubile Nees* [1]), the abundance of fungi, clearly prove that the east-monsoon on Krakatao, as a rule, is repeatedly interrupted by rains.

Mr. Docters van Leeuwen says that near the top he found *already* patches of *Saccharum*. He might better have said that he found *still* patches of this grass, which, together with *Imperata*, is on not too dry soils one of the first settlers in deforested localities. Afterwards, at least when things take their natural course, these two species are gradually superseded by young forest. We may safely assume that 30 years ago *Saccharum* was more common in the higher parts of Krakatao than in 1919, the year in which Mr. Docters van Leeuwen made his trip. In a subsequent paper [2]) the latter affirms that there were on Krakatao still large stretches of *Saccharum* and *Imperata*, which „lent a peculiar aspect to the island". Such large stretches of *Imperata* intermixed with *Saccharum* may be met with everywhere in thinly populated deforested not too dry localities in Java. From the description given of the vegetation in 1919 may be deduced that in the investigated parts the forest was furthest developed and therefore probably oldest in the ravines at an altitude of 600—700 m. This confirms my hypothesis (see pp. 26, seq.) that in these ravines part of the old vegetation has survived the catastrophe and that they have afterwards become starting-points of a young vegetation.

§ 7. On April 27th Mr. Docters van Leeuwen and his companions embarked, rounded the north side of the island and landed at Zwarte Hoek. The steep wall of the rupture was distinctly becoming green by grasses and some ferns growing on it. On the beach at Zwarte Hoek were all sorts of littoral plants, among which *Spinifex littoreus Merr.* Among the thick brushwood stood a few young specimens of Cocos *nucifera L.* and also a well-developed specimen [3]) of *Cycas rumphii Miq.* This species is an exceedingly slow grower. Yet the specimen found cannot date from before the eruption because this beach did not yet exist then. In the excursion of the congress in Sept. 1919 a ♂ specimen was found on the same spot. Ernst who explored the same locality in 1906, did not mention the specimen found by Mr. Docters van Leeuwen, which was certainly not older than 15 years. But Ernst speaks of another specimen which Mr. Docters van Leeuwen was unable to trace.— The visitors did not tarry long here because the cliffs bore only a meagre vegetation. From Zwarte Hoek they went to Verlaten Eiland where they stayed for 2 days. On April 29th they returned home.

[1]) This species seems to be rare in Java. Up to now it has only been found in a few localities in wet places at comparatively low (30—800 m.) altitudes. See hereafter the list of „forest plants" sub No. 64.

[2]) The Vegetation of the Island of Sebesy, situated in the Sunda-strait, near the Islands of the Krakatau group; in the year 1921, Ann. Jard. Bot. Buitenzorg XXXII (1923), 139.

[3]) The Dutch text adds that this was a ♀ specimen.

As stated on p. 178 the steep wall of the rupture bore in 1908 already many plants. The beach at Zwarte Hoek consists, as was already mentioned on p. 145, of 2 oecologically very different parts, the beach proper bearing halophilous plants and the old leached-out parts with a halophobous vegetation. Between these two Mr. Docters van Leeuwen does not distinguish. It would be of interest to compare the vegetation of 1906 and 1908 with that of 1919, but a list of all plants found at Zwarte Hoek in 1919 not having been given, this is impossible. We are only told that there were „all sorts of „littoral plants", a popular statement such as one would scarcely expect to find in a scientific paper on the development of a new vegetation. *Spinifex* was collected at Zwarte Hoek already in 1897 and on most or all subsequent excursions (See pp. 89, 146, 179). We are not told whether the thick brushwood mentioned consisted of littoral plants growing near the floodmark or of non-littoral plants growing farther back. But in this brushwood the only (♀) specimen of *Cycas rumphii* was found that was seen in April 1919. In the Transactions of the Congress no further data are given about the habitat of this specimen but in a subsequent paper [1]) we are informed that this female plant was growing immediately behind the beach-vegetation „near the cone of tumbled stones and masses of „ash" [2]) and that it was quite conspicuous. It (its trunk ?) was 1,20 m. high. Another specimen, a male one, was discovered in October of the same year near the ♀ one; it „flourished more under the trees" and (its trunk ?) was 0,95 m. high. This rather large specimen had apparently been overlooked in April, which proves how carelessly the investigation in that month had been carried out. Both specimens possessed stems rising upright from the base. This is considered by Mr. Docters van Leeuwen proof that the growth of *Cycas* is not slow. *Cycas* is, under cultivation, indeed a very slow grower (see pp. 130, 131) but we know nothing at all about the velocity of its growth in its natural habitat: the sea-shore and the localities not very far from it. Mr. Docters van Leeuwen says [1]) „that it is „difficult to believe that this pair of robust plants should have sprung „from pieces of trunk washed ashore". This unbelief is rather unfounded. The trunk of *Cycas rumphii* very often produces lateral shoots.

---

[1]) Dr. W. Docters van Leeuwen. The vegetation of the Island of Sebesy, situated in Sunda-Strait near the islands of the Krakatau-group; in the year 1921, in Ann. Jard. Bot. Buitenzorg XXXII (1923), p. 138.

[2]) It seems that Mr. Docters van Leeuwen means the large talus of ashes bordering the beach of Zwarte Hoek on its south-eastern extremity.

(see p. 130, footnote 2). From a trunk washed ashore but not yet dead [1]) such a shoot may quite well have developed and of course have grown in an upward direction. The mother-trunk may afterwards have died and rotted away. — In the same brushwood the coco-palms were found, of the apparent age of which nothing is said but that they were young. They cannot have been the same as those seen by me 11 years before, for the latter were growing some hundreds of meters to the north-west. If these had survived they would have been in 1919 rather tall and could have borne flowers and fruits, they would have been very conspicuous and probably have been observed. As not any mention is made of them, it is next to certain that they have disappeared in the intervening years. Why the plants forming the „meagre vegetation of the cliffs" are not recorded is not clear. In short, the description of the vegetation at Zwarte Hoek given by Mr. Docters van Leeuwen is altogether unsatisfactory. It appears that his aim was not to study seriously the development of the vegetation but only to hastily collect as many species as possible.

§ 8. Mr. Docters van Leeuwen states that his investigation was only a provisional one and that he shall have to return several times, when not only the species of the flora and fauna have to be collected as completely as possible but also a picture must be formed of the phenomena that concurred in reestablishing life in these islands.

I am of opinion that, as first of all it has not been proven that the original vegetation was totally destroyed by the eruption of 1883 and secondly the results of all investigations were strongly biassed by their very serious shortcomings, we know at present next to nothing about the *mode of development* of the new flora. And we shall *never* know, if the investigations continue to be carried out in the very unscientific way followed up till now. Collecting some plants and vaguely speculating on them should, in the absence of all serious investigations, not be considered as identical with trying to solve difficult oecological problems. The sooner there comes an end to this childish delusion, the better it is. The money misspent on costly botanical excursions and publications of this kind could be used much better.

§ 9. It is a not unusual phenomenon that somewhere in the ocean a new island arises. This is mostly formed by the action of coral-polyps. Coral-islands are generally small, flat, level islands, which, with the thin layer of sand cast on them by the sea, do not possess enough of fertile soil for forest-plants. The

1) One should bear in mind that the transport of objects by sea-currents from Java to Krakatao can be made within a single day.

plants that manage to live on them at the start, are accordingly, as a rule, those hard starvelings, accustomed to be content with a very modest spot on the earth, which can make a shift to subsist on what little there is of food, and capable of defying the unfavourable conditions of soil and climate; these are generally the littoral and the mangrove-plants. The latter require mostly for their growth an oozy muddy soil, whereas the genuine beach plants will put up even with comparatively bare coral-rocks. As a general rule the first plants are conveyed to these islands by ocean-currents. Among these plants also belongs the coco-palm. In the opinion of Mr. Docters van Leeuwen the study of Krakatao has established it as a fact that the coconut is capable of spreading independently of man. Therefore, on finding a coco-palm sprouting in the driftwood at only a few yards' distance from the sea, he fixed this fact by means of a photo. — After certain pioneers borne by the sea have been washed ashore on the islands, these are visited by birds, which may carry all sorts of seeds attached to their feathers or in their intestines. Moreover the wind also brings seeds and spores.

I cannot agree with the somewhat fantastical description given here of the character and manner of development of the flora on very young coral-islands. I pointed out already (See p. 63) that such islands consist, so to say, entirely of a young beach, hence contain in their soil much NaCl, lack in their centre a basin of sweet water (stored rainwater) and therefore can bear, at least in the east-monsoon, only a halophilous vegetation [1]). There is no reason to suppose that their *lack of fertility* is such as to prevent forest from developing on them. Tropical *forests* do not require a fertile soil though individual *species* may do; in many regions of Java it may be observed that forests thrive quite well in unfertile localities, if only they can dispose of a sufficiently *deep* soil and a *sufficient amount of water.* — But *young* coral islands do not possess a deep soil, neither — at least in the east-monsoon — in that soil a great water-supply, hence they can bear no forests but only scattered trees or not even these. As coral-islands grow older and increase in size, the coral weathers to some depth and they may gradually store rain-water in their centre. Such old coral-islands *are very often densely wooded, provided the trees are spared by man,* which is frequently *not* the case.

That young coral-islands need not at all be unfertile is abundantly shown by the vigorous habit of the herbaceous and shrubby halophytes often copiously growing on them. These are by no means the „hard starvelings" Mr. Docters van Leeuwen speaks so pathetically of, but often quite robust individuals, needing no man's pity.

The edaphic differences between the habitats of mangrove-plants and „genuine beachplants" have not well been grasped by Mr.

---

[1]) In the west-monsoon one often finds on young beaches seedlings of non-halophilous species, but these die off the next east-monsoon or already earlier.

Docters van Leeuwen. Most mangrove-plants do require a deep, muddy soil, but some species f.i. *Sonneratia alba Smith, Rhizophora stylosa Griff., Rhizophora conjugata L., Aegiceras corniculatum Gärtn.* and *Avicennia marina Vierh.* may thrive quite well on almost bare rocks, provided these are daily exposed to the tide. Very many „genuine beach-plants" on the contrary, among them numerous herbaceous ones, do not content themselves with „comparatively bare rocks" but require a rather deep soil, f.i. *Albizzia retusa Bth.* – *Caesalpinia crista L.*, – *Casuarina equisetifolia L.* – *Dodonaea viscosa Jacq.* (the littoral form), – *Euphorbia atoto Forst.*, – *Hernandia peltata Meissn.* – *Josephinia imperatricis Vent.* – *Remirea maritima Aubl.* – *Spinifex littoreus Merr.* – *Tribulus cistoides L.* – *Triumfetta repens Merr. et Rolfe,* – *Vitex ovata Thunb.* and many others.

Mr. Docters van Leeuwen found somewhere on Verlaten Eiland one seedling of Cocos *nucifera* at a few yards' distance from the sea, from which fact he concludes that this species can spread independently of human aid. As it is not certain that the nut had been carried there without any human aid and as moreover it is *not at all sure that the plant will survive*, the seedling, even now that it has been photographed and „this fact has been fixed", can hardly be considered to have furnished the required proof. For such a proof much more unimpeachable arguments are needed. At Zwarte Hoek in 1906 seedlings of this palm were seen, that proved to have disappeared in 1908; in 1908 other seedlings were observed that were not found back in 1919. If the specimens of 1906 and 1908, whether planted or not, were not able to maintain themselves, why should we assume beforehand that the single young seedling of 1919 will be fit to do so. As the exact habitat of the photographed specimen is not indicated, it will be impossible for future investigators to locate it with certainty; it can perhaps not be found back.

For the speculations on the introduction of seeds and fruits in the intestines of birds, I refer to p. 13 and p. 71. As to import of seeds attached to the feathers of birds or carried over by wind, such seeds seem not yet to have been observed on Krakatao. The plants supposed to be issued from them lead, for the present, only an easy theoretical life on paper, not a more difficult real one in Nature.

§ 10. In this paragraph Mr. Docters van Leeuwen gives an excerpt of Treub's paper on Krakatao, whose opinions as to the complete destruction of the old flora and the mode of introduction and development of the new one he fully adheres to. He adds that the pioneer ferns were *only* [1]) such species as were used in their natural homes to „open sunlight and a non-waterretaining soil", such as *Ceropteris calomelanos Und.* and other kinds. Three years after the eruption Krakatao was, at the coast, furnished with littoral plants and more in the interior with algae and ferns. A few grasses and bushes were also to be met with already here and there. Investigations however, have disclosed, that there also are other plants, whose seeds and fruits can float in sea-water for some length of time without losing their germinative power; they also will therefore be washed ashore and can develop as a part of the new strand-flora, but gradually they are killed, choked [2]) or ousted by the real beach plants.

The paper of Treub being already at length discussed in Chapter IV, I need not refute here the general views of Mr. Docters van Leeuwen. His statement that the pioneer ferns were *only* such as are used to „a non-waterretaining soil" *is most incorrect.- Acrostichum aureum L.* [3]), *Stenochlaena palustris Bedd.* [4]) and *Pteris tripartita Sw.* [5]) require a marshy, wet, or at least humid soil; *Ceropteris calomelanos Und.* [6]), *Blechnum orientale L.* [7]) and *Nephrolepis exaltata Schott* [8]) do not thrive in a very dry one. As regards the 3 last-named ferns Mr. Docters van Leeuwen makes here the same mistake as is commonly made by unexperienced collectors; he takes into consideration only the surface of the soil, not the layers in which the plants root. – As to his description of the plant-covering of Krakatao in 1887, he does not for a moment consider that Treub investigated only *a small part* of the lower slopes on the north-western side, so that his conclusions need not hold good for the rest of the mountain. – As to the non-littoral plants whose seeds are washed ashore and germinate there, but which „are gradually killed, choked or ousted „by the real beach-plants", Mr. Docters van Leeuwen quite forgets in this explanation to take into account edaphic factors. The true littoral plants are halophytes and grow in a soil unfit for non-littoral species. As I already observed herebefore one has to carefully distinguish between young, salt-containing beaches and old, leached-out ones. And one should also consider that, where the sea rapidly washes away the land, the border of the latter may not be a beach in the oecological sense of the word, and also that many

[1]) Italics by me. (B).
[2]) The English text has here, probably by misprint, *chocked*. From a comparison with the Dutch text, which uses here the word „*verstikt*", appears that the right reading is such as given above.
[3]) See p. 55, No. 4.
[4]) See p. 54, No. 2.
[5]) See p. 59, No. 10.
[6]) See p. 54, No. 1.
[7]) See p. 55, No. 3.
[8]) See p. 56, No. 6.

seeds of non-littoral plants washed ashore may, after having dried, be driven by the wind to salt-free localities in the interior and germinate there [1]). It is clear that the problem of the settlement of new plants introduced by the sea is much more complex than Mr. Docters van Leeuwen but for a moment supposes. But as all investigations on Krakatao on this subject are entirely lacking, it is needless to further discuss the matter.

§ 11. In this paragraph the excursion of 1897 is discussed. Penzig and his companions found the flora much enriched and, what was most remarkable, plant-communities (of which term Mr. Docters van Leeuwen gives a definition) had already been formed. Penzig distinguished 3 formations: that of the beach-plants, the *Barringtonia*-formation and that of the grassy wilds.

Penzig's paper was already discussed in Chapter VI to which I refer. As to the first formation of plant-communities, see p. 90. Mr. Docters van Leeuwen's nomenclature of the strand-formations is somewhat obscure. He distinguishes between beach-plants and the *Barringtonia*-formation, forgetting that the latter also consists exclusively of beach-plants. This *Barringtonia*-formation may either border immediately on the sea or be secluded from it by the *Pes caprae*-formation. This depends on the nature of the beach; on decreasing beaches the *Pes caprae*-formation is very often lacking whilst on increasing sandy [2]) beaches it is as a rule, present. It seems that by the name of beach-plants Mr. Docters van Leeuwen meant to indicate only the *Pes caprae*-formation. Besides the 3 formations mentioned by Mr. Docters van Leeuwen, Penzig found a fourth, the (kremnophytic) vegetation [3]) of ferns growing on the steep sides of the rocks. — The *Barringtonia*-formation was found by Penzig only on Verlaten Eiland: it consisted of a grove of young specimens of *Casuarina equisetifolia L.* [4]), 5—6 m. high. Four years afterwards Haeckel [5]) observed on the coast of Krakatao a small grove which seemed to reach a height of 10—15 m. As to the components of this grove one can only guess. I shall refrain from doing so.

§ 12. In 1906 Ernst and his companions visited Krakatao. The formation of plant-communities had made further progress; the number of plants had vastly increased. Ernst records for the 3 expeditions together 108 species [6]) for the whole Krakatao-group. The majority of these plants belonged to the

1) See p. 12, footnote 1.
2) Increasing *clayey* beaches bear, as a rule, mangrove-forest.
3) See p. 31.
4) See pp. 125 and 205.
5) See p. 27, footnote 3.
6) This is indeed the total number of Phanerogams and ferns recorded by Ernst, in which the finds of Valeton (See Chapter VII) are not included. After the necessary corrections have been made, this number of 108 is reduced to 96.

common beach-plants whose seeds are conveyed by ocean-currents. The list of plants given by Ernst contains under plants found *near*[1]) the beach a number of species generally not met with in these places, but after some time when the beach-community and the *Barringtonia*-belt shall have expanded, those others will be pushed more inwards or become extinct. The additions to the flora observed in 1906 consisted mainly of Phanerogams; the Cryptogams, among them the ferns, had not greatly increased. The grassy wilds still occurred but the isolated clusters of trees were already more numerous than in 1897. In the ravines were already found a few forest-trees, but, not having succeeded in ascending the mountain, Ernst did not know what things looked like higher up.

The paper of Ernst was discussed in Chapter VIII. Penzig and Ernst each investigated only a very small part of the island. Moreover, Ernst and his companions collected very many of their plants on the south-eastern side of the island, which was never visited by Penzig; contrariwise they did not examine the basaltic rocks and the slopes above Zwarte Hoek where Penzig gathered his inland plants. Hence it is quite impossible to make conclusions as to the additions of the flora between 1897 and 1906 or to modifications of the inland-flora. Of the 83 species[2]) of Phanerogams recorded by Ernst for the entire group no more than 40[3]), hence *not a majority but a slight minority*, have seeds or fruits which, as a rule (or at least very often), are conveyed by ocean-currents. Such plants as Ernst found *near* the beach, i.e. in (partly at least) salt-free localities, are very frequently found in such spots, it is quite incorrect to say that they generally are not met with in these places. The prediction of Mr. Docters van Leeuwen as to their being *pushed inward* in future by littoral species is a baseless speculation. They may in course of time be killed by the all-invading forest of the interior or by a decrease of the beach, but this is not what Docters van Leeuwen means. He does not distinguish between true littoral plants, which are halophytes, and plants growing on leached-out parts of old beaches and localities behind

1) Italics by me (B.).
2) After the necessary corrections have been made.
3) Viz.: *Barringtonia asiatica Kurz*, — *Caesalpinia crista L.*, — *Calophyllum inophyllum L.*, — *Canavalia rosea D.C.*, — *Cassytha filiformis L.*, — *Casuarina equisetifolia Forst.*, — *Cerbera manghas L.*, — *Clerodendron inerme Gärtn.*, — *Colubrina asiatica Brongn.*, — *Cycas rumphii Miq.*, — *Derris heterophylla (Willd) Backer*, — *Desmodium umbellatum D.C.*, — *Dodonaea viscosa Jacq.*, — *Entada phaseoloides Merr.*, — *Erythrina variegata L.* var. *orientalis Merr.*, — *Euphorbia atoto Forst.*, — *Guettarda speciosa L.*, — *Hernandia peltata Meissn.*, — *Hibiscus tiliaceus L.*, — *Indigofera zollingeriana Miq.*, — *Ipomoea denticulata Chois.*, — *Ipomoea longiflora R. Br.*, — *Ipomoea pes caprae Sw.*, — *Ischaemum muticum L.*, — *Morinda citrifolia L.*, — *Pandanus tectorius Sol.*, — *Pithecolobium umbellatum Bth.*, — *Pongamia pinnata Merr.*, — *Premna integrifolia L.*, — *Remirea maritima Aubl.*, — *Scaevola frutescens Krause*, — *Sophora tomentosa L.*, — *Spinifex littoreus Merr.*, — *Terminalia catappa L.*, — *Tournefortia argentea Vent.*, — *Vigna marina Merr.*, — *Vitex paniculata Lamk.*, — *Wedelia biflora D.C.*, — *Ximenia americana L.*, — *Xylocarpus granatum Koen.*

it, which are quite able to hold their own there against littoral species, as may be observed everywhere on the Javanese coasts. Not the littoral *species* but the *waves of the sea* may become the enemies of inland plants. The few „*forest-trees*" named by Ernst were *not* found in the ravines but immediately behind the *Casuarina*-forest and in open places in the grass-jungle. – The dense forest in the ravines was by Ernst and his companions only *seen from afar* but could not be *reached* then. Two years afterwards some of the lower parts of this forest were investigated as is described in Chapter IX.

§ 13. In this paragraph the expedition of 1908 is discussed.

For this expedition see Chapter IX. According to Mr. Docters van Leeuwen I stated in my paper [1]) that behind the littoral belt first a region of mixed forest came, then the grass-jungle. This I did not at all say, I expressly stated that the belt of mixed forest was only present here and there and that it was quite narrow. In the Dutch text of this paragraph it is incorrectly stated that Mr. Demmeni *during* my visit reached the top. As a matter of fact he did so a few days after I had left the island.

§ 14. In 1919 much appeared to have been altered, even after making due allowance for the fact that this time the summit was reached. For also in the lower ravines a great many more plants occurred than formerly. The list of Mr. Docters van Leeuwen, which contains also the previous finds, numbers 272 [2]) plants for the entire group, against 127 in the list of Ernst [3]). The beach-plants had not greatly increased: of genuine littoral plants only seven new ones had made their appearance. This belt of the flora is the best explored because one is each time obliged to traverse [4]) it when one wishes to reach the other zones Contrarily the number of forest-plants has been greatly added to. Mr. Docters van Leeuwen, when confining his list of forest-plants to Phanerogams and ferns, found 68 species in all, whereas Ernst and Backer mention 20 species. The highest rate of increase was shown by vascular Cryptogams of which Mr. Docters van Leeuwen collected 45 species, which means about double the number of those in 1906. Ernst has one epiphyte growing near the coast whereas Mr. Docters van Leeuwen

---

1) Jaarverslag van den Topographischen Dienst in Ned.-Indië (Year-book of the Topographical Survey Service in the Dutch Indies), 1908, p. 190. See also herebefore p. 187.

2) According to the *Dutch* text 277 species. For Krakatao itself, after the necessary corrections and additions have been made, 242 species are mentioned, of which 135 Phanerogams, 43 vascular Cryptogams, 6 mosses (3 of them found by Mr. Docters van Leeuwen on this trip and recorded by Max Fleischer in Ann. Jard. Bot. Buitenzorg XXXIII (1923), p. 105 seq.), 1 lichen, 34 fungi and 23 other Cryptogams.

3) According to the Dutch text 137 species. The total number of species found on Krakatao on and before the excursion of Ernst is 119, of which 76 were Phanerogams, 14 vascular Cryptogams, 3 mosses, 3 fungi and 23 other Cryptogams.

4) The Dutch text has the same reading: „Deze strook van de flora wordt na-„tuurlijk het beste onderzocht, omdat men er telkenkere weer door moet, wanneer men „de andere gordels wil bereiken".

found 16 forms. Genuine fungi appear in Ernst's list to the number of 3; Mr. Docters van Leeuwen found 33 of them. Most noteworthy is the great increase of epiphytes; many of them demanding other surroundings than other plants. Many species desire a constantly moist atmosphere and therefore thrive most luxuriantly in the humid wild forests of the mountains. At about 600 m. [1]) *Trichomanes humile* was found, whose leaves are very unfit to resist a long desiccation. A great change in the conditions of life must have come about and this was that the few forest-plants which were to be found in the ravines during Backer's visit have multiplied astonishingly since 1908, covering the ravines with a dense forest, consisting of few species but of an incredible number of individuals of each. In 1919 they already formed deep shadows and shed leaves plentifully, the ground was covered with a rather thick [2]) layer of leaf-mould, and with the humus arises the possibility of the development of numerous forest-plants. Moreover the mountain-peak is frequently wrapped in clouds, consequently very moist (as Mr. Docters van Leeuwen had occasion to be aware when climbing it) and also so a good habitat for epiphytes.

Mr. Docters van Leeuwen is wrong in saying that during my visit in 1908 but few forest-plants occurred in the ravines. The lower parts of the „waldartige Bestände" observed there as early as 1906 by Ernst and his companions were reached in 1908, and in my short paper on the expedition of that year (See Chapter IX) I expressly stated that, though the number of *species* found was small, the number of *individuals* belonging to them was often enormous. As a matter of fact, many of the ravines investigated in 1908 at an altitude of 200–400 m. were densely wooded; as we came higher up the mountain the trees (of the same species) became larger, from which may be inferred that the woods had extended from above downwards. Most probably this process has continued ever since 1908; in 1919 the forests may have approached considerably nearer to the coast than in 1908. The woods between 200 and 400 m., still comparatively young in 1908, were in 1919 eleven years older, the trees were higher, their crowns bigger and denser, consequently the forest darker, more humid, better fit for epiphytes; the undergrowth may have become richer in species. This is the ordinary course of events with young secondary forests in rainy climates in Java (See p. 209). *Cyrtandra sulcata Bl.* which in 1908 was not *seen* [3]) beneath an altitude of 250 m. was now already collected at 50 m. Hence Mr. Docters van Leeuwen may be right in saying that in the lower ravines much more plants occurred than formerly, though this is not proven and, moreover,

[1]) According to the Dutch text. The English text gives 1800 feet or ± 540 m.
[2]) This is the correct translation of the Dutch text: vrij dik, which is somewhat vaguely rendered in the English text by: „of some thickness".
[3]) This is not the same as *did not occur*.

it is most improbable that the ravines examined in 1919 were the same as those of 1908. Unhappily Mr. Docters van Leeuwen does not tell *which* plants he found beneath 400 m. and *where* and in *what numbers* he found them, so that his remark is of little value for future investigations. The few data contained in Appendix III to his paper are altogether unsufficient. As nothing is known of the flora of the ravines above 400 m. in 1908 and Mr. Docters van Leeuwen gives but very scanty and unsatisfying details of the vegetation observed by him in the higher parts of the mountain, it is not only quite impossible to compare the flora of 1919 with that of former years, but in later years it will be very difficult to draw reliable conclusions about changes in the vegetation after 1919. Nothing is known with certainty.

The flora of the beach is indeed, as a rule, the best explored, not because one has to traverse it at every visit, which in most places is a work of a few seconds or at best a few minutes, but because, when the tide is low, *the beach itself forms a quite convenient and inviting road for the exploration of this narrow belt of the vegetation.* Nevertheless rare or inconspicuous species may be quite well overlooked on the south-eastern and eastern beaches of Krakatao which have a length of about $3^3/_4$ km. and are densely clothed with trees, shrubs and herbs. *A really scrupulous investigation would require several weeks.* Mr. Docters van Leeuwen asserts that 7 (the number given in the Dutch text is 10) *new genuine littoral* plants have made their appearance since 1908. He does not give the names of these but on consulting Appendix I, one finds mentioned 8 beach-plants and halophytes not yet recorded for former excursions, which of course by no means proves that they were not yet present then, the beach having been never yet thoroughly investigated. Only one of these, *Mucuna gigantea D.C.* [1]), was found on Krakatao. This is a rather large climber not rarely met with in the *Barringtonia*-formation. The seven other species were found on Lang Eiland and Verlaten Eiland. Only two of them (*Monerma repens P. B.* and *Thuarea involuta R. Br.*) are genuine beach-plants; *Halophila ovalis Hook.*, a marine species, inhabits, as a rule such coral terraces adjacent to the beach as run dry or almost so, when the tide runs low but are

[1]) The English text wrongly mentions this species as found by Ernst on Verlaten Eiland. Ernst quite rightly recorded his plant under the name of *Mucuna pruriens D. C.* In the Dutch text the specimen of Krakatao (1919) is mentioned as *M. pruriens D. C.* The carelessness wherewith Mr. Docters van Leeuwen has written his paper, the many discrepancies between the English and the Dutch texts, make it a desperate task to search out anything reliable and valuable in it.

inundated during the flood. The four other halophytes (*Excoecaria agallocha L., Lumnitzera racemosa Willd., Paspalum vaginatum Swartz* (= P. distichum Auct.) and *Sesuvium portulacastrum L.*) are no genuine beach-plants but mere inhabitants of a saline, clayey, humid or (in the case of *Paspalum*) even inundated soil, irrespective of this being found quite near the sea or not; all of them may be found in localities far remote from the coast, they are no **genuine** beach-plants. But Mr. Docters van Leeuwen apparently does not distinguish between genuine littoral plants and halophytes.

As to forest-plants, one wishes Mr. Docters van Leeuwen had told what to understand by this popular term. There are many kinds of forests in the tropics and in most forests several oecological groups of plants may be distinguished. Among the 68 „forest-plants" [1]) recorded in Appendix III (of which 22 were already found at former excursions) there are 10 of which *he himself* says that they are *no real forestplants.* But why then were they inserted in his list? And what is the difference between a forest-plant which is a *real* forest-plant and a forest-plant which is *not* a *real* one? In my discussion of Appendix III I shall revert to this curious list.

As to Vascular Cryptogams, after the necessary corrections have been made [2]), the number found by Mr. Docters van Leeuwen is reduced to 40 which number rises to 44 when the species found on previous trips but not collected in April 1919 are included. This is rather more than about double the number of those found in 1906. For Ernst and his companions (See p. 145) found in 1906 not more than 5 vascular Cryptogams, whilst the list of Ernst containing all previous finds, mentions but 14 species after the necessary corrections have been made. *Ernst did not find a single fern living as an epiphyte;* he expressly states: „An den Felsen sammeln wir auch .....*Drynaria quercifolia J. Sm.* (= Polypodium quercifolium L.), einen der wenigen Epiphyten der *Barringtonia*wälder, *der hier aber,* ähnlich wie in der Umgebung der Solfataren und Krater der Vulkane Javas, *in Felsspalten wächst*" [3]). But Ernst did not reach the wooded ravines in which in April 1919 most of the epiphytes were found.

[1]) The Dutch text says 69 species, but Appendix I to the Dutch text records only 68 species.

[2]) *Dryopteris calcarata* O.K. (See p. 56, No. 7); *Dryopteris flaccida* O.K. (See p. 57, No. 8); *Dryopteris vilis* O.K. (being but a form of Dryopteris setigera O.K.); *Dryopteris setigera* O.K. var. *Backeri* V. A. v. R. (differing not at all from Dr. setigera O.K.) and *Nephrolepis tomentosa* V. A. v. R (being but a form of N. exaltata Schott) should be expunged from the list. *Lycopodium cernuum L.* and *L. squarrosum Forst.* are in the Dutch text recorded for the trips of 1919, in the English text of Appendix I not.

[3]) See Ernst, Neue Flora Vulkaninsel Krakatau (1907), pp. 36, 37. For *Drynaria quercifolia L.* see p. 113, No. 26.

If by genuine Fungi Mr. Docters van Leeuwen means *Eumycetes* – and what else can he mean? – he did not find 33 [1]) but 32 species on the main island (38 on the 3 islands together), as recorded by himself in Appendix I. Including previous finds this number rises to 34 (40).

The increase of the number of epiphytes – mostly mosses, ferns and orchids – in a young forest growing older is in a wet tropical climate a quite common phenomenon. There is nothing remarkable about the fact that this also happened on Krakatao unless it can be proven that these epiphytes were introduced after the eruption and light is spread upon the manner how this happened. Many of the epiphytes and fungi found for the first time in 1919 may already have been present *in other parts of the island* in 1908, even long before that year. On this point also nothing is known with certainty. All statements about a great change in the conditions of life after 1908 on Krakatao, *taken as a whole*, are mere guesses, unsupported by facts.

§ 15. The Krakatao-flora can be divided into the following spheres. First the usual littoral zone which already began to form in 1886 and was well represented, in 1897. Next the *Barringtonia*-formation which is now wider, now narrower, sometimes absent. Then the *Casuarina equisetifolia*-forest with its peculiar vegetation of smaller trees, shrubs and creepers, to which Backer applied the name of mixed forest. After this we get to the grassy wilds of the level interior which occasionally also rises to the ridges. Here are found groves chiefly consisting of various species of *Ficus*, of *Pipturus incanus Wedd.*, – *Macaranga tanarius M. A.* and other trees and shrubs. These clusters of trees become more and more numerous, would seem to coalesce and with time will occupy the greater part of the steppe. At present the photo still shows them isolated. Here and there this grassy plain advances as far as the littoral zone. This is succeeded by the young virgin forest of the ravines, stretching as far as the top. In some parts there are still dry ridges set with ferns and *Saccharum spontaneum* and at the summit one comes upon the last remains of the fern-zone, mingled with *Saccharum* and stunted trees from the ravine-flora. This covers and will probably continue to clothe for some time to come, the summit proper as well as the very highest ridges. Continued research will be needed to define all this more completely. The most remarkable thing is that especially in the higher ravines the forest is formed by practically one plant, *Cyrtandra sulcata Bl.*, which in Java and Sumatra is a small forest-plant under tall trees. But this plant also will probably after a lapse of time have to be content with a more modest place.

To this there is but little to add. By the usual littoral zone Mr. Docters van Leeuwen apparently means only the *Pes caprae*-formation, forgetting that the *Barringtonia*-formation and *Casuarina equisetifolia* belong quite as well to the beach in the oecological sense of the word. The name of mixed forest was *never* applied by me to the *Casuarina*-forest with its undergrowth. For this subject

[1]) The same number is given in the Dutch text.

see p. 187. The „grassy wilds", as a whole, are not level but slope upwards towards the interior of the island, as may be seen from the picture given by Mr. Docters van Leeuwen [1]) himself. In 1908 they covered very many of the ridges. If in 1919 they really only „occasionally rose to them", why then did Mr. Docters van Leeuwen not tell us with which other plants these ridges were clothed in that year? One wishes Mr. Docters van Leeuwen had told which trees, besides the few mentioned ones, were components of the groves in the grass-jungle. Those found in 1908 are recorded in Chapter IX. Now it is impossible to make a comparison between the finds of 1919 and those of 1908.

As it is not at all proven that on the south-eastern side and on the top of Krakatao there ever existed a fern-zone, the ferns found on the summit need not be considered as remains of this quite hypothetical belt. Finally it should be stated that in Java *Cyrtandra sulcata Bl.* is not a small forest-plant but a rather robust shrub, often of $1^1/_2$ to 3 m. high.

§. 16 The two main problems to be solved in studying the new flora of the totally devastated islands of the Krakatao-group are, firstly: In what manner did the plants get there? and secondly: In what way was the soil made suitable to yield a sufficiency of food to the plants? The researches on the second of these problems have been very inadequate and now it is too late: for a long time the ground has ceased being sterile as it was shortly after the eruption. Volcanic matter will mostly weather readily, especially when this is furthered by the action of lower organisms. From the analyses of ashes and pumice made by Winkler and published by Verbeek, appeared that nearly all the elements were present, except nitrogen. On the beach this could be disengaged from organic matter washed ashore. Nitrogen-containing acids were brought to the soil in rain-water. De Kruyff found nitrogen-binding bacteria in the soil-samples brought back by Ernst. Algae, bacteria and moulds play their parts in this process, being organisms which, like the *Cyanophyceae* found by Treub, are extremely sober in their requirements of nutriment. These little plants by their growth cause the surface-soil to crumble and moulder more easily and furthermore, when they die and decay, small amounts of organic matter enter the soil, which are transformed again by other micro-organisms. Eventually such soils will sustain a few frugal plants. On the beach the usual littoral plants found what was sufficient for them in the way of nutrition; typical representatives of this formation throve here after a short time and this contributed to enable other plants start living a few years later, a little more in the interior [2]). These were mainly ferns, the same kinds as content themselves in other places also with dry, slightly weathered soils. The beginning therefore was principally the transportation of such soil-organisms as are

[1]) Ann. Jard. Bot. Buitenzorg XXXI (1921), tab XIX, beneath.

[2]) The Dutch text has here quite another reading, viz.: „en bovendien konden na „enkele jaren in het binnenland andere planten komen", i.e. „and moreover, after a few „years, other plants could arrive in the interior".

wont to effect the chemical transformation in the earth, which were succeeded first by ferns and beach-plants and next after by grasses that are organized to live on dry exposed soils, such as *Saccharum spontaneum L.* and *Imperata cylindrica P.B.* By their agency the amount of organic substance was gradually added to, rendering the soil increasingly suitable for the growth of other plants. In this connection it is remarkable that among the pioneers were the orchids *Spathoglottis plicata Bl.* and *Arundina speciosa Bl.* which in Java frequently grow on dry, steep banks. Orchids are among those plants that live in symbiosis with fungi in their roots; these fungi are often indispensable even to their germination. Not only must the seeds of those orchids have got to Krakatao in great numbers but also the fungi. A large number of tropical plants, not only forest-plants but beach-plants as well, as Dr. Von Faber has told in his lecture of Oct. 3d 1919, live in symbiosis with fungi. Therefore an abundant microflora is necessary for the life of the higher plants.— The number of bacteria per gramme of earth was in the samples brought by Ernst from Krakatao almost equal to that in the soil of Buitenzorg. Several *Leguminosae* have been found with their nitrogen-tubercles; these plants help energetically in rendering the soil fertile.

Many plants could not grow on Krakatao before the humidity of the surroundings had become greater. This only became possible when the pioneers of the arboreal flora, especially species of *Ficus,* had begun to build up forests in the ravines. In these ravines watery vapours clung more easily and longer than in the flat open ground; a layer of humus was formed, suitable to smaller forest plants, the vegetation also becoming gradually so dense that a moist shade prevailed beneath the trees, just the thing for numerous forest-plants, especially epiphytes. In this humus vegetate all kinds of fungi preparing the conditions of life for higher saprophytes, that live in community with fungi. Several orchids might be classed among these [1]). *Loranthaceae* are entirely absent. This fact should be accounted for from the absence of the birds that disseminate the *Loranthus*-seeds.

For a discussion of the contents of the first part of this paragraph (for the greater part unavowedly borrowed from Ernst's paper [2]), I refer to p. 73. Herebefore I have already argued that it is not proven, not even probable, that the soil of Krakatao was completely sterilized by the eruption of 1883. Contrariwise, as stated on p. 74, an effect of the heating of the soil may have been a partial sterilisation, causing an increased activity of nitrogen-fixing and nitrification organisms.— Like all his predecessors Mr. Docters van Leeuwen forgets that the old woods covered by the eruptive products formed an enormous store of organic compounds. Wherever the ashes and pumice were soon washed away or nearly so, the organic material came on or near the surface again and could be-

[1]) The Dutch text adds here: „and probably also the species of *Schizaea*, a kind „of fern belonging to the hemi-saprophytes". This statement is probably due to confusion with *Psilotum*. I do not know that it has ever been proven that *Sch. dichotoma Sm.* (the only species of the genus found till 1919 on Krakatao) lives in community with fungi. Dr. Von Faber, the well-known Buitenzorg specialist to whom I applied for information, kindly told me by letter that he knew nothing about the occurrence of a fungus in *Schizaea*-roots.

[2]) Ernst, Neue Flora Vulkaninsel Krakatau (1907), pp. 51 seq.

come a source of food for the new vegetation. (See also p. 85). There is not a single indication that this new vegetation developed earlier on the shore than in the interior, hence it will be difficult to prove that the littoral vegetation more than any other one „contributed to enable other plants to start living a few years later, „a little more in the interior". Unless – and this is a most alluring hypothesis – it did so from a distance by some hitherto unknown kind of emanation.

The mistake made by Mr. Docters van Leeuwen in considering all ferns found by Treub as plants content with dry soils, was already discussed on p. 221. Such another mistake is made by him where he says that *Saccharum spontaneum L., Spathoglottis plicata Bl.* and *Arundina speciosa Bl.* are plants of dry soils. As a matter of fact these plants, especially the first, require rather much moisture to prosper (See pp. 100 seq.). On p. 221 I already pointed out that Mr. Docters van Leeuwen takes here into regard only the dry surface of the soil seen by him, not the moist deeper layers. As to the large number of beach-plants living in symbiosis with fungi, Mr. Docters van Leeuwen has apparently not well understood the statement of Dr. Von Faber who did not speak of *beach-plants* but of *plants of coral-islands* [1]), which is quite another thing. In the leached-out interior of the *older* coral-islands a lot of non-littoral species live. Dr. Von Faber kindly informed me that such beach-plants as are halophytes do *not* possess mycorrhizae. Among the others which grow on a non-saline soil there are some which have mycorrhizae, others which live in symbiosis with bacteria. – Considering that *Leguminosae* on Krakatao up to 1919 have been found only on the beach and the localities immediately behind it (See p. 170), it is difficult to conceive how these plants could help *„energetically"* in rendering the soil fertile. *In his list of „forest-plants"* (Appendix III) *Mr. Docters van Leeuwen does not record a single Leguminosa.*

It is quite true that many plants could not live on Krakatao before there were humid forests. Such forests, or at least groves, may in the higher ravines have been present within comparatively few years after the eruption. Treub when approaching Krakatao *saw as early as 1886 rather large plants growing near the top.* The plants requiring much shade and moisture which were found on

[1]) Handelingen Eerste Nederlandsch Indisch Natuurwetenschappelijk Congres (1919), p. 128.

Krakatao in 1908 and afterwards, may either have survived the eruption in the form of seeds, rhizomes or spores or have been introduced from elsewhere; possibly both cases have presented themselves. It is very probable that most, if not all, epiphytic plants were killed by the eruption; if their seeds or spores have not survived (what nobody knows) they must have been introduced in some way or other. It is very tempting to ascribe this introduction to direct transport by the wind but as yet there exist no solid proofs for this hypothesis; other ways are possible, though perhaps less probable. Mr. Docters van Leeuwen does not give the names of the „several orchids" which might be classed among the higher saprophytes. All species recorded in his list of orchids in Appendix I possess normal green leaves, consequently are no saprophytes or at best hemi-saprophytes [1]). This can be ascertained only by an examination of the roots, which seems not yet to have been carried out. It has, to my knowledge, not yet been proven that the seeds of the terrestrial orchids *Spathoglottis plicata Bl.* and *Arundina speciosa Bl.* in their natural habitat cannot germinate without the aid of fungi. Neither has it been proven that the seeds of these orchids have been carried over to Krakatao *in great numbers;* this is but a bold hypothesis. Both species are frequently autogamous [2]), hence the numerous specimens found in 1908 on Krakatao may quite well have originated from a single specimen, even from a single seed of each. Nor is anything proven about the *supposed* transport of the fungi which Mr. Docters van Leeuwen *supposes* to live in symbiosis with them. The *Loranthaceae* wich have been present at Krakatao before 1883 (see p. 16) were certainly killed by the eruption or died soon afterwards, though seeds may have survived. Though it is not proven that in 1919 they had not yet reappeared — only a small part of the island having been investigated — we may believe that if they were present at all, they were still far from common. The explanation of this fact given by Mr. Docters van Leeuwen may be correct.

§ 17. In this paragraph Mr. Docters van Leeuwen discusses the mode of transport of plants to Krakatao. Leaving man out of account, and at first he of course contributed little or nothing, seeds and plants may be transported by wind, by ocean-currents and by birds and bats. All these have done their share. Treub and Penzig found chiefly such plants as were conveyed by the sea and the winds. Penzig only mentions 4 species whose seeds were probably conveyed across by birds. This is natural when one reflects that the barren naked island offered nothing in any way alluring to the birds

[1]) The only saprophytic orchid found till 1919 on Krakatao is *Eulophia macrorrhiza Bl.* (see p. 180) which is *not* mentioned by Mr. Docters van Leeuwen.

[2]) See p. 108.

because there was no shade and little food. Plants, whose distribution is undoubtedly effected by birds, were already found by Ernst. Mr. Docters van Leeuwen also found fairly great numbers of such plants. But still most of the new plants are even now brought by winds and sea. It is not feasible to state the exact way in which each plant arrived because comparatively few positive data are available to us and many kinds of plants may be distributed in more ways than one. In the first appendix to his paper Mr. Docters van Leeuwen has stated the manner of spreading with the nearest approach to accuracy he could, but often he has been compelled to guess [1]).

Besides these two main questions (treated of in §§ 16 and 17) there are a few of less importance. One of these, the spread of Cocos *nucifera*, Mr. Docters van Leeuwen has already touched upon in passing. Another point inviting investigation was to him the presence of a great number of galls. This subject was separately discussed by him [2]).

As I have already repeatedly stated, nothing at all is known of the vegetation or the fauna of the higher parts of Krakatao before 1908, nor of the manner in which the plants possibly introduced into the interior after the eruption, have arrived there. We do not even know whether any plants were survivors of the eruption. There is no *proof* that birds have really introduced plants. Of course they *may* have done so, but nobody knows it with certainty. It is not *proven* that the wind has carried fruits, seeds or spores to Krakatao. This *may* have happened, — especially with the minute spores of the Cryptogams and seeds of orchids it is highly probable — but there has been made not a single experiment, not even a single observation in corroboration of this statement. As Mr. Docters van Leeuwen correctly says, it is *impossible* to state the manner and way of introduction. Nevertheless he has in Appendix I to his paper freely ventured forth guesses on this point. Such guesses do not in any way advance science. On Cocos *nucifera* I set forth my views already on p. 39 herebefore. Of the (afterwards to be discussed) paper on galls referred to by Mr. Docters van Leeuwen, I will at present only say that it contains nothing that brings us nearer to a solution of the problem. Some galls have been found and in some way described, but the life-history of the animals living in them was not studied. No observations as to modes of transport have been made, but only remarks borrowed from others or vague speculations without any scientific value.

§ 18. In this final paragraph Mr. Docters van Leeuwen makes some remarks on the fauna, for the greater part admittedly borrowed from the paper of

---

[1]) The Dutch text has here quite another reading: „ik heb bij benadering de ver- „spreidingswijze aangegeven, doch ik heb er dikwijls een slag naar moeten slaan". i. e.: „I „have roughly stated the manner of spreading but I was often compelled to make „hazardous guesses".

[2]) Dr. W. Docters van Leeuwen. The Galls of Krakatau and Verlaten Eiland in 1919, in Ann. Jard. Bot. Buitenzorg XXXI (1921), p. 57.

Mr. Jacobson [1]) on this subject. For the study of the development of the new *flora* these remarks are of no importance. It should only be noted that Jacobson in 1908 observed no bats on Krakatao during the 3 nights he passed there. Neither did he find earth-worms (which were found in 1919, see p. 221).

In 1908 Mr. Jacobson did not reach the older parts of the higher ravines in which earth-worms may have been present. The investigation in 1908 was altogether unsufficient to warrant a statement as to the non-occurrence of such small animals as are easily overlooked.

Of the 5 appendices to the paper of Mr. Docters van Leeuwen I shall first discuss Appendix II giving a list of plants found *on* and *near* the beach, hence containing oecologically very different plants. This list contains 86 species, but omitting those not found on the main island and reducing one as a synonym to another species we arrive at a total of 67. These are the following, arranged after the system of Christensen (Ferns) and Engler and Prantl. About each species not yet formerly found on Krakatao I have given a few data on its distribution on Java and its means of reproduction. About the distribution in South-Sumatra no sufficient data are extant.

PLANTS FOUND ON AND NEAR THE BEACH.

Schizophyllaceae.

1 (86). [2]) *Schizophyllum alneum Schröter* (=Sch. commune Fries).

Cosmopolitic fungus, living on decaying wood and such-like substrata, spread throughout the East-Indian Achipelago. The reproduction is effected by spores.

No littoral plant. This fungus was found on bamboo-stems on the beach. We are not told whether these had been washed ashore or had been imported by Mr. Händl for the construction of his buildings (See p. 196). Living bamboos had as late as 1919 not yet been collected in the island.

Polypodiaceae.

2 (81). *Nephrolepis biserrata Schott.*

Terrestrial or epiphytical fern, rather common throughout Java from the plains up to an altitude of ± 1000 m. except in regions where the east-monsoon is very strong. It grows

[1]) Edw. R. Jacobson, De nieuwe fauna van Krakatau, in Jaarverslag van den Topographischen Dienst in Nederlandsch Indië (Annual of the Topographical Survey Service in the Dutch Indies) 1908, 192.

[2]) The number within brackets is that given in Appendix II to Mr. Docters van Leeuwen's paper.

by preference in more or less shadowed localities, in rain-forests either primary or secondary, in teak-forests, in jungles, in and around native villages. The reproduction is brought about by spores and by long creeping stolons. It often grows gregariously.

No littoral plant. As appears from the Dutch text, collected in the wood of *Casuarina equisetifolia* near the beach.

3 (82, 83). *Nephrolepis exaltata Schott.* (Recorded in Appendix I and II under the names of *N. hirsutula Presl* and *N. tomentosa V. A. v. R.*) See p. 56, No. 6 and p. 212.

No littoral plant. As appears from the Dutch text, specimens were collected in the wood of *Casuarina equisetifolia* near the beach and on rocks near the sea. See also hereafter the list of forest-plants sub No. 7.

4 (85). *Stenochlaena palustris Bedd.* See p. 54, No. 2.

No true littoral plant, though it may by found quite near the beach. In the English text wrongly recorded as a true beach-plant; in the older Dutch text rightly mentioned as a species not belonging to the typical littoral flora. As appears from the Dutch text, specimens were collected in the wood of *Casuarina equisetifolia* near the sea.

5 (78). *Ceropteris calomelanos Und.* See p. 54, No. 1.

No littoral plant. Habitat not mentioned.

6 (84). *Pteris vittata L.* (Recorded in Appendix I and II under the name of *Pt. longifolia L.*) See p. 56, No. 5.

No littoral plant. As appears from the Dutch text, specimens were collected on rocks near the sea.

7 (79). *Cyclophorus acrostichoïdes Presl.*

Epiphytical or very rarely terrestrial fern, spread throughout Java from the plains up to an altitude of 700 m., in sunny or moderately shadowed localities. The reproduction is effected by spores and by ramification of the long creeping rhizomes.

No littoral plant. Found on a specimen of *Casuarina equisetifolia* near the beach.

8 (80). *Drynaria quercifolia J. Sm.* See p. 113, No. 26.

No littoral plant. As appears from the Dutch text, collected on a specimen of *Casuarina equisetifolia* near the beach.

9 (77). *Acrostichum aureum L.* See p. 55, No. 4.

In Appendix II wrongly recorded as a true beach-plant. It is often found miles away from the sea, frequently on the inner border of a broad mangrove-belt, rather rarely in sweet water.

Habitat not mentioned.

Cycadaceae.

10 (76). *Cycas rumphii Miq.* See p. 130, No. 7, and p. 217.

In the English text wrongly recorded as a true beach-plant, in the older Dutch text rightly mentioned as a species not belonging to the typical littoral flora.

In § 7 mentioned as only found at Zwarte Hoek.

Pandanaceae.

11 (75). *Pandanus tectorius Sol.* See p. 43, sub No. 4.

Habitat not mentioned. But from a subsequent paper [1]) it appears that this species was observed in the old forest of *Casuarina equisetifolia*, i.e. on the south-eastern side of the island.

Gramineae.

12 (63). *Imperata cylindrica P. B.* (I. arundinacea Cyr.). See p. 106 and § § 6 and 16 herebefore.

No littoral plant. Mentioned for the interior only.

13 (69). *Saccharum spontaneum L.* See p. 100 and § § 4, 5, 6, 15 and 16 herebefore.

No littoral plant. On the south-eastern side behind the *Casuarinas.*

14 (68). *Pogonatherum paniceum Hack.* (P. crinitum Trin.). See p. 105.

No littoral plant. As appears from the Dutch text collected on steep slopes near the sea.

15 (64). *Ischaemum muticum L.* See pp. 89, 90.

In § 4 mentioned as found on the south-eastern side.

16 (61). *Eriochloa ramosa O. K.*

Grass of periodically humid or swampy, but in the east-monsoon frequently very dry, rather heavy soils. — This American plant, which was introduced in Java already long ago,

---

[1]) Dr. W. Docters van Leeuwen, The Vegetation of the Island of Sebesy, situated in the Sunda Strait, near the Islands of the Krakatau-group; in the year 1921, in Ann. Jard. Bot. Buitenzorg XXXII (1923), p. 146.

seems to have been found there only beneath an altitude of 75 m. In the western part of the island it has been collected in few localities only, but in the drier central and eastern parts it is rather common. The reproduction is effected by seeds which are copiously produced.

Possibly the specimen collected by Mr. Docters van Leeuwen (habitat not mentioned) has been wrongly named. It might be *E. subglabra Hitchc.*, likewise of American origin, which is quite common in West-Java, especially on moist or swampy heavy soils. This grass has, to my knowledge, never yet been found fruiting in Java but seems to be reproduced exclusively by fragments of the culms which readily take root.

Habitat not mentioned. Neither *E. ramosa* nor *E. subglabra* are littoral plants though both may be found near the sea.

17 (66). *Hoplismenus* [1]) *compositus P. B.*

Grass with creeping and copiously rooting culms, occurring throughout Java from the plains up to an altitude of ± 2000 m. in shadowed, not too dry localities, on road-sides under trees and along not too dark forest-paths. The reproduction is brought about by ramification of the culms and by the fruits which, when ripe, remain enclosed within the spikelets. These are provided with a rather thick awn which becomes viscous when the fruit ripens [2]) and attaches itself easily to the coat of animals or to garments, by which expedient the entire spikelet is carried away.

No littoral plant. As appears from the Dutch text collected in the wood of *Casuarina equisetifolia* near the sea.

18 (70). *Spinifex littoreus Merr.* See pp. 89, 90 and §§ 4 and 7 herebefore.

On the south-eastern side and at Zwarte Hoek.

19 (62). *Eleusine indica Gärtn.*

Strongly tillering grass, much spread throughout the cultural zones of Java from the plains up to an altitude of ± 2000 m., occurring as a weed in fields, gardens and lawns and especially along much trodden roads. The reproduction is effected exclusively by fruits which are produced in great numbers and germinate easily.

No littoral plant. Habitat not mentioned.

---

1) The generic name is frequently written *Oplismenus*.

2) The same case presents itself with the very closely allied *H. undulatifolius P.B.*, which apparently is very rare in Java. See i.a. Hegi, Illustrierte Flora von Mittel-Europa I, 190.

Cyperaceae.

20 (56). Cyperus *cylindrostachyus Boeck.* (recorded in Appendix I and II as C. *cyperoides* O.K.).

Perennial weed, spread throughout Java from the plains up to an altitude of ± 1800 m., in shadowed localities. The reproduction is brought about by fruits.

No littoral plant. Habitat not mentioned.

21 (55). Cyperus *pennatus Lamk.* See p. 47, sub No. 5; p. 90; p. 91, No. 5.

Wrongly recorded in Appendix II as a true beach-plant. It may be found quite near the beach or even on it, but it is much more frequent in humid clayey localities behind it, even far remote from the sea. It occurs as often in non-saline localities as in saline ones.

Habitat not mentioned.

22 (58). *Kyllinga monocephala Rottb.*

Perennial weed, spread throughout Java from the plains up to an altitude of 1200 m., especially beneath 300 m., in moderately shadowed or sunny, by preference grassy localities, very often under lawn-trees and on road-sides. The reproduction is effected by ramification of the creeping rhizomes and by fruits.

No littoral plant. Collected in the wood of Casuarina *equisetifolia* near the beach.

23 (57). *Fimbristylis spathacea Roth* (F. glomerata Nees). See p. 47, No. 6; p. 90; p. 91, No. 5.

In the English text wrongly recorded as a true beach-plant. Though often occurring in humid saline localities and frequently growing gregariously there, this species occurs but rarely on the beach but is much more frequent in clayey soils behind and often far remote from it.

Habitat not mentioned.

24 (59). *Remiria maritima Aubl.* See p. 47 and p. 132, No. 12.

Habitat not mentioned.

Palmae.

25 (74). Cocos *nucifera L.* See p. 39, No. 3 and § § 4 and 7 herebefore.

On the south-eastern side and at Zwarte Hoek.

Dioscoreaceae.

26 (60). *Dioscorea* spec.

None of the Javanese species of this genus is a littoral plant, though *D. bulbifera L.* sometimes – though rarely – may be found on sandy beaches. This species is reproduced by its winged seeds and by axillary bulbils which easily fall off after having reached a somewhat large size. On the beach of one of the coral islands near Batavia Kamerling [1]) found a living bulbil of a species of *Dioscorea* washed ashore.

A specimen of some species of *Dioscorea* was collected in the *Casuarina equisetifolia* forest near the beach.

Orchidaceae.

27 (73). *Spathoglottis plicata Bl.* See p. 107 and §§ 5 and 16.

No littoral plant. As appears from the Dutch text, collected „near the sea".

Casuarinaceae.

28 (7). *Casuarina equisetifolia L.* See p. 91, No. 6 and § 4 herebefore.

On the south-eastern side.

Urticaceae.

29 (49). *Pipturus incanus Wedd.* See p. 134, No. 19 and §§ 5 and 15 herebefore.

No littoral plant. As appears from the Dutch text, collected in the wood of *Casuarina equisetifolia* near the sea. As appears from § 5 herebefore the species was found in 1919 not only near the sea but also at some distance from it in a dry water-channel.

Aristolochiaceae.

30 (3). *Aristolochia tagala Cham.*

Twiner spread throughout Java from the plains up to an altitude of 800 m. in brushwood and secondary forests, always in scattered specimens though sometimes locally not rare. The reproduction is effected by the winged seeds, which may be easily spread by water and perhaps over some small distance over land also by wind.

---

[1]) Natuurkundig Tijdschrift Ned.-Indië LXXXI (1921), p. 31.

No littoral plant. Collected in the forest of *Casuarina equisetifolia* near the beach.

Amarantaceae.

31 (2). *Cyathula prostrata Bl.* (C. geniculata Lour.).

Perennial weed with ascending, copiously rooting stems spread throughout Java from the plains up to an altitude of ± 1600 m. in shadowed localities, especially in places where people or animals frequently move along. The erect inflorescences consist of a rather long main axis bearing glomerules of 2–3 ☿ flowers surrounded by a number of abortive flowers transformed into hooks. The indehiscent fruit remains enclosed within the perianth. When the fruits of a glomerule are all ripe the latter easily lets loose as a whole from the main axis; the hooks frequently cling to garments or to the coat of animals, which by this expedient may convey the seeds.

No littoral plant. Habitat not mentioned.

Menispermaceae.

32 (33). *Tinospora coriacea Beumée* (T. crispa Diels).

Twining dioecious weed, or shrub, spread throughout the lower regions of Java in hedges, thickets and light forests. The reproduction is effected by the bright red drupes containing each a single kernel. It is not known whether these drupes are eaten by animals.

No littoral plant. Collected in the *Casuarina equisetifolia* forest near the beach.

Lauraceae.

33 (28). *Cassytha filiformis L.* See p. 91, No. 7 and § 4 herebefore.

On the south-eastern side.

Leguminosae.

34 (34). *Pithecolobium umbellatum Bth.* See p. 147, No. 20.

Wrongly given in Appendix II as a true beach-plant. Habitat not mentioned.

35 (39). *Desmodium umbellatum D. C.* See p. 148, No. 22.

Habitat not mentioned.

36 (38). *Derris heterophylla Backer* (D. uliginosa Bth.).

Wrongly recorded in Appendix II as a true beach-plant. It may occur on the beach but is much more frequent along tidal creeks and in the mangrove, often far remote from the sea.

Habitat not mentioned.

37 (40). *Erythrina variegata L.* var. *orientalis Merr.* (E. indica Lamk.). See p. 46, sub No. 1.

Habitat not mentioned.

38 (37). *Canavalia rosea D.C.* (In Mr. Docters van Leeuwen's paper mentioned as C. lineata D. C.) See p. 89, p. 92, No. 8 and § 4 herebefore.

On the south-eastern side.

39 (42). *Vigna marina Merr.* See p. 89, p. 93, No. 9 and § 4 herebefore.

On the south-eastern side.

Euphorbiaceae.

40 (25). *Macaranga tanarius M. A.* See p. 135, No. 26 and §§ 5 and 15 herebefore.

No littoral plant. As appears from Appendix II to the Dutch text, collected in the forest of *Casuarina equisetifolia* and according to §§ 5 and 15 also growing in the grassy wilds of the interior.

41 (23). *Euphorbia atoto Forst.* See pp. 90, 93.

Habitat not mentioned.

Vitaceae.

42 (53). *Columella trifolia Merr.* (Vitis trifolia L.). See p. 136, No. 28 and § 4 herebefore.

Wrongly recorded in Appendix II as a true beach-plant. The glabrous variety mostly occurs in the less salt parts of the mangrove which often extend to some kilometers from the beach. The pubescent variety is found to far in the interior. From the paper of Mr. Docters van Leeuwen it does not appear which form he found.

Malvaceae.

43 (30). *Hibiscus tiliaceus L.* See p. 93, No. 11.

Habitat not mentioned. But from a subsequent paper [1])

[1]) W. Docters van Leeuwen, The Vegetation of the Island of Sebesy, situated in the Sunda-Strait, near the Islands of the Krakatau-group; in the year 1921, in Ann. Jard. Bot. Buitenzorg, XXXII (1923), p. 146.

it appears that this species occurred in the forest of *Casuarina equisetifolia.*

Guttiferae.

44 (27). *Calophyllum inophyllum L.* See p. 44, No. 6.
Habitat not mentioned.

Caricaceae.

45 (6). *Carica papaya L.*

Short-living fruit-tree, extensively cultivated throughout Java and in waste places and light forests sometimes subspontaneous. The large berries, which are comparatively cheap, are often consumed by the natives. Before eating, the fruit is cut into parts and the numerous seeds which clothe the wall of the large central cavity are cast away.

In 1906 Ernst and his companions found many apparently spontaneously growing specimens on Verlaten Eiland near the coast; in 1908 I myself found numerous specimens on Lang Eiland and Verlaten Eiland likewise near the coast; in 1919 specimens were collected on Krakatao in the *Casuarina equisetifolia* forest near the beach.

No littoral plant.

Lecythidaceae.

46 (29). *Barringtonia asiatica Kurz* (= B. speciosa Forst.). See p. 93, No. 12.

Habitat not mentioned. But from a subsequent paper [1]) it appears that this species occurred in the old forest of *Casuarina equisetifolia.*

Combretaceae.

47 (9). *Terminalia catappa L.* See p. 39, No. 2.

Habitat not mentioned. But from a subsequent paper [1]) it appears that this species occurred in the old forest of *Casuarina equisetifolia.*

Melastomataceae.

48 (32). *Melastoma malabathricum L.* See p. 150, No. 33.

No littoral plant.

---

[1]) W. Docters van Leeuwen, The Vegetation of the Island of Sebesy, situated in the Sunda-Strait, near the Islands of the Krakatau-group; in the year 1921, in Ann. Jard. Bot. Buitenzorg, XXXII (1923), p. 146.

As appears from the Dutch text, collected „to close to „the sea, especially on rocks".

Convolvulaceae.

49 (18). *Ipomoea denticulata Chois.* See p. 157, No. 27 and § 4 herebefore.

On the south-eastern side.

50 (19). *Ipomoea longiflora R. Br.* (I. grandiflora Lamk.). See p. 157, No. 28 and § 4 herebefore.

On the south-eastern side.

51 (20). *Ipomoea pes caprae Sw.* See p. 48, No. 7 and § 4 herebefore.

On the south-eastern side. Wrongly recorded as a true beach-plant. This species very frequently occurs far from the sea.

Borraginaceae.

52 (4). *Tournefortia argentea L. fil.* See p. 54, No. 8.

Habitat not mentioned.

Verbenaceae.

53 (51). *Lantana camara L.*

Shrub, originating from tropical America, already more than sixty years ago purposely introduced into Java where it spread with marvellous rapidity and soon became one of the commonest plants, as it is to this very day. It grows from the plains up to an altitude of ± 1650 m. in pastures, hedges, jungles and secondary forests and very frequently invades deserted tea- and coffee plantations. The reproduction is effected by the small drupes which possess a hard kernel containing two seeds. The fruits are often eaten by native children, moreover by several kinds of birds and by a mammal *(Paradoxurus hermaphroditus Schreb.)* in whose droppings one frequently finds the undamaged kernels.

No littoral plant. Habitat not mentioned.

54 (52). *Premna integrifolia L.* See p. 137, No. 37.

Wrongly recorded in Appendix II as a true beach-plant. Though very frequently found on the beach, the species often occurs in the plains far from the beach in non-saline localities.

Habitat not mentioned.

55 (50). *Clerodendron inerme Gärtn.* See p. 151, No. 37.

Wrongly recorded in Appendix II as a true beach-plant. Though frequently found on the beach the species much more often occurs along tidal creeks and other water-sides, not rarely rather far from the sea.

Habitat not mentioned.

Rubiaceae.

56 (46). *Oldenlandia subulata Korth.*

Annual of periodically dry sandy or stony soils, rare in Java. The reproduction is effected by seeds which are produced in great numbers and possess floating power. They are easily spread by floating rain-water.

No littoral plant. Collected on rocks near the sea.

I have not seen the specimen gathered by Mr. Docters van Leeuwen. It may have been wrongly named; it might be O. *corymbosa L.* which in 1908 was found at the base of the basaltic rocks at Zwarte Hoek. This latter species is not recorded in the lists given by Mr. Docters van Leeuwen.

57 (44). *Guettarda speciosa L.* See p 138, No. 41.

Habitat not mentioned. But from a subsequent paper [1]) it appears that this species occurred in the old forest of *Casuarina equïsetifolia*.

58 (45). *Morinda citrifolia L.* See p. 93, No. 16.

Habitat not mentioned. But from a subsequent paper [1]), it appears that this species occurred in the old forest of *Casuarina equisetifolia*.

Cucurbitaceae.

59 (21). *Trichosanthes bracteata Voigt.* (In Appendix II to the English text and in Appendix I and II to the Dutch text recorded under the name of *Tr. tricuspidata*. It is dubious whether *Tr. bracteata Voigt* specifically differs from *Tr. tricuspidata Lour.*). See p. 139, No. 43.

---

[1]) W. Docters van Leeuwen, The Vegetation of the Island of Sebesy situated in the Sunda-Strait, near the Islands of the Krakatau-group; in the year 1921, in Ann. Jard. Bot. Buitenzorg, XXXII (1923), p. 146.

No littoral plant. As appears from the Dutch text, found climbing in *Casuarinas* near the beach. In Appendix II to the Dutch text wrongly recorded as not yet mentioned in the list given by E r n s t.

Goodeniaceae.

60 (26). *Scaevola frutescens Krause.* See p. 49, No. 9.
Habitat not mentioned.

Compositae.

61 (16). *Vernonia cinerea Less.* See p. 118, No. 58.
No littoral plant. As appears from the Dutch text collected in the forest of *Casuarina equisetifolia* near the sea.

62 (10). *Blumea lacera D. C.* See p. 51.
No littoral plant. Habitat not mentioned.

63 (14). *Pluchea indica Less.* See pp. 50, 108, 119.
Wrongly recorded in Appendix II as a true beach-plant. The plant never occurs on the beach proper but very often on clayey or calcareous soils behind it.
Habitat not mentioned.

64 (17). *Wedelia biflora D.C.* See p. 49, under No. 1.
Habitat not mentioned.

65 (12). *Eleutheranthera ruderalis Sch. Bip.*
Annual ruderal weed, originating from tropical America, introduced into Java already long ago, at present spread throughout the lower regions of the island, especially in the western and central parts up to an altitude of ± 700 m. in not too dry localities. It is a common weed of fields, road-sides and village-groves, in the moister regions growing everywhere, in the drier ones especially in shadowed places and there, locally, often very common. The reproduction is effected by the fruits, which are produced in large numbers and possess no pappus; they are easily spread by the tillage of the soil and by water flowing down.
No littoral plant. Habitat not mentioned.

66 (15). *Synedrella nodiflora Gärtn.*
Ruderal weed, originating from tropical America, introduced into Java already long ago, at present spread throughout the island from the plains up to an altitude of ± 1200 m.,

especially in moderately shadowed localities, under hedges, on road-sides, in native villages, in coffee-plantations. It is reproduced exclusively by fruits which lack a pappus and are curiously dimorphous, those of the outer flowers being surrounded by a broad, deeply serrate wing, those of the disk-flowers being much narrower and provided with 2–4 spreading apical awns. The fruits are spread by water flowing down and by animals to whose coat they attach themselves.

No littoral plant. Collected as appears from the Dutch text in the *Casuarina equisetifolia* belt near the sea.

67 (13). *Erechthites hieracifolia Raf.* See p. 52, under No. 4.
No littoral plant. Habitat not mentioned.

The above list of 67 species collected on the main island contains under the head „Plants observed by me (i.e. Mr. Docters van Leeuwen) on or near the beach" very heterogeneous elements: plants of sandy beaches, mangrove-plants, simple halophytes and plants of salt-free localities in the interior (which are often found quite close to the sea). The latter plants belong to divers oecological groups, such as kremnophytes, weeds of fields and road-sides, common jungle-plants and meso-hygrophytes. Mr. Docters van Leeuwen distinguishes only between „true beach plants" or „genuine „littoral plants" and „plants that should not be considered to belong „to the beach flora proper". The term „true beach plant" or „genuine „littoral plant" should, in my opinion, be restricted to plants which occur exclusively, or almost so, on the beach, i. e. on the narrow zone immediately behind the flood-mark. Beach-plants should not be confounded with mangrove-plants nor with simple halophytes, which may grow on the beach but occur also far behind it and even quite in the interior. But Mr. Docters van Leeuwen evidently comprises under „true beach plants" all halophytes either growing usually on the beach or not. As a true beach plant he even records in the English text No. 4 (85), *Stenochlaena palustris Bedd.*, which is neither a halophyte nor a real beach plant (though it may be found in the *Barringtonia*-formation) and No. 63 (14), *Pluchea indica Less.*, a shrub of heavy soils either saline (and very often so) or not. Though most frequent in the coastal plains the latter species may be also found far in the interior but it never occurs on the beach proper. Of many species the habitat has not been given; it is not even said whether they were collected at Zwarte Hoek or on the eastern side of the island. Of not a single species the frequency is mentioned. Hence it is impossible to draw from this most unsatisfactory list conclusions as to possible modifications of the littoral flora since the time of former excursions, nor will anyone be able to do so in future years. It can only be stated that in 1919 not a single littoral plant (with the exception perhaps of *Mucuna gigantea D. C.*, for which see beneath) was found on Krakatao that had not already been collected in 1908 or former years. No less than six or seven littoral species or mangrove-components found at previous excursions (*Albizzia retusa Bth.*, – *Dodonaea viscosa Jacq.* [1]) – *Indigofera zollingeriana Miq.*, – *Sophora tomentosa L.*, – *Thespesia populnea Soland.*, – *Ximenia americana L.* and *Xylocarpus granatum Koen.*) were in 1919 not found back or at least not collected on the main island. This, of course, is by no means proof that they were not present then, as even large plants may be easily overlooked by unsufficiently trained and careless collectors. Of seven other littoral plants or halophytes which according to Appendix I of the Dutch and English texts were collected on Krakatao during the short trip made in October 1919 by the members of the Congress to that island, but not in April of that year, no less than 5 (*Cerbera manghas L.* – *Hernandia peltata Meissn.*, – *Mucuna gigantea D. C.* [2]), – *Pongamia pinnata Merr.* and *Vitex paniculata Lamk.* [3]) are not recorded in Appendix II, though it can hardly be believed that they were not growing on or quite near the beach. Perhaps Mr. Docters van Leeuwen, though recording them in Appendix I as having been found on Krakatao, excluded them from Appendix II because he did not know them to be beach-plants. But this cannot be considered a valid reason, because information on this point was easily available. The two other species (*Acrostichum aureum L.* and *Morinda citrifolia L.*) are recorded in Appendix II (No. 77 and 45).

---

1) In a subsequent paper [Dr. W. Docters van Leeuwen, The Vegetation of the Island of Sebesy, situated in the Sunda-strait, near the Islands of the Krakatau-group; in the year 1921 in Ann. Jard. Bot. Buitenzorg XXXII (1923), p. 146] it is affirmed however that this species occurred (when?) in the forest of *Casuarina equisetifolia*.

2) In the Dutch text the name is given as *Mucuna pruriens D. C.*, which is an inland plant, found but rarely on the beach and only on its older parts. It was collected in 1906 on Verlaten Eiland and recorded by Ernst. In the English text of Mr. Docters van Leeuwen's paper the name of the specimen found in 1919 has, without any comment, been changed into *M. gigantea D. C.*, which is a plant of the beach and the localities immediately behind it. In this same text the specimen found in 1906 on Verlaten Eiland is also – and wrongly – reduced to *M. gigantea D.C.* It is a difficult task to unravel the mess made by Mr. Docters van Leeuwen.

3) See p. 183, No. 41.

Summarizing, from the paper of Mr. Docters van Leeuwen we do not learn whether the littoral vegetation in 1919 differed from that of 1906 or 1908. Much less has it been proven that this vegetation had grown richer in species in the 11 or 13 intervening years. The investigation of the littoral flora which presents no serious difficulties but requires much care, was carried out in April 1919 so very negligently that half a year afterwards no less than seven species, *all of them large perennials,* were collected that had not been found in April, though at least 5 of them were already present in 1906 or 1908. There is no reason to believe that these species had died out in April and had reestablished themselves again in October. Even a big (male) specimen of the very conspicuous *Cycas rumphii Miq.* with a trunk of almost one meter high was overlooked on the beach of Zwarte Hoek (Cf. § 7). Of many of the plants found *near* the shore for the first time in 1919 we do not know on which side of the island they occurred. The researches on the coastal flora of 1919 have given us nothing reliable to build upon, *they have in no way advanced our knowledge of the development of the flora.*

In Appendix III to his paper Mr. Docters van Leeuwen has recorded 68 species under the head: „Forest-plants found by me on Krakatau". Herebeneath these species have been arranged in accordance with the systems of Christensen (Ferns) and Engler and Prantl. Of each species not yet formerly found on Krakatao I have given a few data on its distribution over Java and its means of reproduction. Of none of the species Mr. Docters van Leeuwen has mentioned the frequency; of many the habitat and the vertical range are not given, the height mentioned apparently being often only that of the locality where a specimen was collected. This appears i.a. from No. 44 of the list, *Ficus fistulosa Reinw.,* of which in § 5 herebefore is said that it was frequent in the lower ravines, whilst in Appendix III it is only given for the „summit of Mount Rakata". In § 5 it is said that *Villebrunea rubescens* was seen in the valleys beneath an altitude [1]) of ± 100 m. (at which point „the way „was blocked by an absolutely unscalable precipice"), in Appendix III the only height given is 200 m. Sometimes (No. 12, 35, 37, 46, 50, 58 of the list beneath) the English text it entirely silent as to the altitude

[1]) In giving altitudes above sea-level I have everywhere followed the original Dutch text. In the English text one meter, which is 3,28 English feet, has been wrongly reckoned for 3 feet.

of the locality where a species was collected, whilst the Dutch text gives information on that point; in such cases I have supplied the missing data from the Dutch text. It also occurs that the Dutch and English text flatly contradict each other. See f.i. No. 3, *Dryopteris megaphylla* C. *Chr.*, which according to the Dutch text occurred up to an altitude of 400 m., but for which the English text gives an altitude of 0–60 m. See also No. 38, *Liparis viridiflora Lindl.* for which species the Dutch text gives a height of 700 m. (about 2300 feet), the English text 208 feet or somewhat more than 60 m. These discrepancies cannot be explained by an error of the press. Analogous but less strongly marked cases are presented by No. 62, *Radermachera glandulosa Miq.*, for which the Dutch text gives: between 30 and 500 m. above sealevel, the English text: between 150 and 1500 feet, (i.e. between 45 and 450 m.) and *Homalanthus populnea O.K.* which according to Appendix III occurred between 30 and 200 m. above sea-level, whilst from a subsequent paper (See No. 57 hereafter) it appears that in April 1919 this species was also collected at an altitude of ± 300 m. Such incongruities clearly show how negligently the paper was composed.

Herebeneath I give firstly the list of „forest-plants", secondly some further remarks on it.

FOREST-PLANTS FOUND BY MR. DOCTERS VAN LEEUWEN ON KRAKATAO.

Hymenophyllaceae.

1 (42) [1]. *Trichomanes humile Forst.*

Minute fern, found throughout Java between 150 and 800 m. above sea-level in constantly moist and well-shaded localities, mostly on wet rocks and steep earth-walls, but also as an epiphyte on the lower parts of tree-trunks.

Collected, as an epiphyte, at an altitude of 600 m.

Cyatheaceae.

2 (43). *Alsophila* spec.

Several species of this hopelessly intricate genus of tree-ferns occur in the moister regions of Java from the lowlands up to high in the mountains. They grow in constantly humid localities, especially in primary and secondary rain-forests and on watersides, often gregariously in ravines but also in

[1]) The number within brackets is that given in Appendix III.

level places. The most common Javanese species seems to be *A. glauca J. Sm.* which is spread throughout the moister regions of the island from the plains up to an altitude of ± 1800 m., especially above 600 m. The reproduction of all Javanese tree-ferns is exclusively effected by spores.

Collected at an altitude of 400 m. In 1908 Demmeni had already seen 3 specimens of a tree-fern (genus unknown) at an altitude of 600 m. Before 1919 the son of Mr. Händl (Cf. Chapter X, p. 197) had observed tree-ferns at a height of 600–700 m.

Polypodiaceae.

3 (52). *Dryopteris megaphylla C. Chr.*

Terrestrial fern with a short rhizome spread throughout the moister parts of West- and Central Java between 250 and 1500 m., in jungles, grassy wilds and secondary forests. The reproduction is effected by spores.

According to the Dutch text occurring up to an altitude of 400 m., but according to the English text from 0–200 feet (± 60 m.).

4 (53). *Dryopteris setigera O.K.* See p. 57, sub No. 8.

Collected in the lower ravines.

5 (45). *Aspidium melanocaulon Bl.*

Kremnophytic fern with a short rhizome, spread throughout the moister regions of Java between 250 and 1000 m. above sea-level, on humid, well-shadowed steep banks, especially on water-sides and in hollow roads. The reproduction is effected by spores.

Collected in the lower ravines.

6 (58). *Nephrolepis biserrata Schott.* See p. 234, No. 2 (81).

Collected in the lower ravines.

7 (59). *Nephrolepis exaltata Schott.* (= N. hirsutula Presl). See p. 56, No. 6, further also § 6 and p. 235, No. 3 (82,83) herebefore.

Collected „from the beach to the summit". The Dutch text adds: „but mostly in little shadowed localities".

8 (49). *Davallia* sp.

The genus *Davallia,* possessing long creeping rhizomes, is in Java from the plains up to high in the mountains represented by several species living either as epiphytes in more or less shaded localities or as kremnophytes.

An epiphytic species of this genus was collected at an altitude of 700 m.

9 (60). *Odontosoria chinensis J. Sm.*

Kremnophytic fern with a short creeping rhizome, spread throughout Java between 600 and 1800 m. above sea-level, especially on steep banks and on sides of hollow roads. The reproduction is effected by spores.

Collected at an altitude of 400 m. on steep slopes in the forest. „*No real forest plant*" (D. v. L.).

10 (50). *Diplazium polypodioides Bl.* (= D. asperum Bl., under which name the plant is recorded in the paper of Mr. Docters van Leeuwen).

Robust terrestrial fern with a thick erect rhizome, occurring throughout Java, except in regions where the east-monsoon is strong, from the plains up to ± 2200 m. above sea-level in shadowed not too dry localities, often in forests, secondary as well as primary ones, on forest-borders, furthermore on well-shadowed road-sides, along water-courses, in hedges and thickets. The reproduction is effected by spores.

Collected at an altitude of 500 m.

11 (47). *Asplenium nidus L.* See § 5 herebefore.

According to the Dutch text collected at an altitude of 100 m.; according to the English text from 0 to 90 m.

12 (46). *Blechnum orientale L.* See p. 55, No. 3.

As appears from the Dutch text collected in the lower parts of the island. The English text adds: „*No real forest-plant* (Treub 1886)"; the Dutch text (translated): „Treub 1886, Penzig 1897, *the forest did not yet exist then*".

13 (44). *Antrophyum* sp. [1]).

Several species of this genus inhabit the moister regions of Java. They grow in densely shadowed, not too dry localities either as epiphytes on tree-trunks or as kremnophytes on steep rock-sides. The reproduction is effected by spores.

---

[1]) In Appendix I to the Dutch text it is said that the species found was either *A. callifolium Bl.* or *A. semicostatum Bl.*

An epiphytic species of this genus was collected at an altitude of 600 m.

14 (51). *Drymoglossum heterophyllum C. Chr.*

Epiphytical fern, spread from the plains up to an altitude of 1000 m., rather common in West- and Central-Java, seemingly rare in the eastern part of the island. It grows on trunks and branches of trees and shrubs whose crowns give but a moderate amount of shadow; the plant needs rather much light. It is often found on trees planted along roads, in gardens and in native villages. The reproduction is effected by ramification of the rhizome and by spores.

Collected at an altitude of 100 m. *„No real forest-plant"* (D. v. L.).

15 (54). *Hymenolepis brachystachys J. Sm.*

Epiphytical fern, possibly only a lowland form of the following species, spread throughout the moister regions of West- and Central-Java from the plains up to an altitude of ± 900 m., mostly on trees planted along road-sides and in native villages, sometimes also in teak-forests. The reproduction is effected by spores.

Collected at an altitude of 500 m.

16 (55). *Hymenolepis spicata Presl.*

Epiphytical fern, spread throughout the more humid parts of Java from the plains up to an altitude of 2800 m., especially above 500 m. It grows in more or less shadowed localities, on trees planted along roads and in native villages, in secondary and primary forests, in tea- and Cinchona-plantations. The reproduction is effected by spores.

Collected at an altitude of 700 m.

17 (61). *Polypodium accedens Bl.* (Pleopeltis accedens Moore).

Small fern with a long creeping rhizome, spread throughout West- and Central-Java between 600 and 1500 m. above sea-level but only in regions where the east-monsoon is feeble. It inhabits well-shadowed humid localities where it lives either on rocks or — and such is mostly the case — as an epiphyte on trees, often on small ones or on rather thin branches. The reproduction is effected by ramification of the rhizome or by spores.

Collected at an altitude of 600 m.

18 (62). *Polypodium heracleum Knze.* (Pleopeltis heraclea V. A. v. R.)

Very robust epiphytical or less often terrestrial fern forming gigantic humus-collecting „nests", spread throughout the moister regions of Java between 600 and 1800 m. above sea-level. The plant grows by preference in moist forests, occasionally on shadowed road-sides.

Collected as an epiphyte at an altitude of 700 m.

19. (63). *Polypodium phymatodes L.* (Pleopeltis phymatodes Moore). See p. 189, No. 4.

Collected in the lower ravines.

20 (64). *Polypodium punctatum Sw.* (Pleopeltis punctata Bedd.). See p. 189, No. 5.

Collected in the lower ravines and at an altitude of 400 m.

21 (65). *Polypodium* (Pleopeltis) spec.

According to the Dutch text collected in the lower parts and at an altitude of 400 m. The English text gives no habitat.

22 (48). *Cyclophorus adnascens Desv.*

Epiphytical fern, spread throughout Java from the plains up to the lower mountainous zone except in regions where the east-monsoon is very strong. It grows by preference in sunny or but moderately shadowed localities, often on trees planted along road-sides and in and around native villages.

Collected at an altitude of 300 m.

Schizaeaceae.

23 (66). *Schizaea dichotoma Sm.*

Terrestrial fern, occurring here and there in West- en Central-Java from the plains up to an altitude of 1000 m. in not too moist parts of rain-forests, in teak-forests and in bamboo-plantations. The reproduction is effected by spores.

Collected at an altitude of 600 m. See also p. 230, footnote 1.

24 (56). *Lygodium circinatum Sw.*

Rather robust twining terrestrial fern spread throughout Java from the plains up to an altitude of ± 1500 m. except in regions where the east-monsoon is very strong. It grows by preference in shadowed not too dry localities, in thickets and hedges, in secondary forests and on wood-borders. The reproduction is effected by spores.

Collected at an altitude of 300 m. According to Appendix I to the Dutch and English texts (where this plant and the next one are transferred to the *Polypodiaceae*) the correctness of the determination was liable to doubt.

25 (57). *Lygodium scandens Sw.*

Rather robust twining terrestrial fern, from the plains up to an altitude of ± 1200 m. spread throughout the moister parts of Java, in thickets, hedges and young secondary forests. The reproduction is effected by spores.

Collected in the lower ravines. „*No real forest plant*" (D. v. L.).

Marattiaceae.

26 (68). *Angiopteris evecta Hoffm.*

Gigantic terrestrial polymorphic fern-species, on account of very unimportant characters and often with the aid of herbarium-fragments only, broken up by authors in a host of minor species that nobody, not even they themselves, can distinguish with certainty. The plant is spread throughout Java between 250 and 1800 m. above sea-level in humid, often densely shadowed ravines, especially along or near water-courses. The reproduction is effected by spores and very often by buds which develop on the big stipular appendices at the base of the petiole after these appendices have fallen off from the short thick trunk [1]).

Collected in the lower ravines.

Ophioglossaceae.

27 (67). *Ophioglossum reticulatum L.* (O. pedunculosum Desv.; recorded in Appendix III to the Dutch text as O. pediculosum Denv.). See pp. 109, 113.

Collected at an altitude of 600 m. The Dutch text adds: „Penzig 1897, when the forest did not yet exist". The English text omits this addition.

1) See R a c i b o r s k i, Ueber die vegetative Vermehrung der Marattiacea *Angiopteris evecta* in Bulletin de l'Académie des Sciences de Cracovie, Classe des Sciences mathématiques et naturelles (1902), p. 48. – Dr. W. D o c t e r s v a n L e e u w e n, Ueber die vegetative Vermehrung von *Angiopteris evecta Hoffm.*, in Ann. Jard. Bot. Buitenzorg XXV (1912), p. 202.

Equisetaceae.

28 (39) *Equisetum debile Roxb.*

Horse-tail, spread throughout the moister regions of Java between 300 and 2700 m. above sea-level, in sunny or moderately shadowed humid or swampy localities, in pools and dry river-beds but also on steep banks and in stony or rather unfertile moist soils, often growing gregariously. The reproduction is effected by ramification of the long creeping rhizomes and by spores.

Collected at an altitude of 700 m. The Dutch text adds: „on a slope bearing but a scanty vegetation". The English text omits this addition.

Lycopodiaceae.

29 (40). *Lycopodium squarrosum Forst.*

Epiphytical or less often terrestrial very variable club-moss, spread throughout the moister parts of Java between 400 and 1600 m. above sea-level, especially in well-shadowed localities, in primary and secondary forests, rarely in stony localites bearing but a scanty vegetation. The reproduction is effected by spores.

An epiphytic specimen was collected at an altitude of 500 m.

Selaginellaceae.

30 (41). *Selaginella plana Hieron.* See p. 190, No. 8.

Collected at an altitude of 500 m.

Zingiberaceae.

31 (38). *Costus sericeus Bl.* See p. 190, No. 12.

Collected from 50 to 200 m. above sea-level. *„No real „forest plant"* (D. v. L.). The Dutch text says: „No real pri-„meval-forest plant".

Orchidaceae.

32 (30). *Habenaria sp.*

Several species of this terrestrial genus of orchids inhabit the moister regions of Java from the lowlands to high in the mountains. They grow either in forests or in grassy plains. The reproduction is effected by seeds.

A species of this genus was collected at an altitude of ± 700 m.

33 (36). *Peristylus gracilis Bl.*

Terrestrial orchid spread in West-Java between 500 and1200 m. above sea-level. The reproduction is effected by seeds.

Collected at an altitude of 700 m.

34 (33). *Nervilia aragoana Gaud.* See p. 191, No. 13.

Collected at an altitude of 100 m.

35 (29). *Eulophia macrostachya Lindl.*

Terrestrial orchid with well developed green leaves, spread over West-Java from the plains up to an altitude of ± 700 m., in thickets, secondary forests and bamboo-plantations. The reproduction is effected by seeds.

Collected, as appears from the Dutch text, at an altitude of ± 50 m. In Appendix I and III wrongly recorded for the expedition of 1908; the species found in that year was the saprophytical leafless *E. macrorrhiza Bl.* See p. 180, No. 16.

36 (34). *Oberonia monstruosa Lindl.*

Epiphytical orchid found in a few localities in the lower mountainous zone of West- and Central-Java. The reproduction is effected by seeds.

Collected at an altitude of 700 m.

37 (35). *Oberonia* spec.

According to the Dutch text collected at an altitude of 700 m.

38 (32). *Liparis viridiflora Lindl.*

Epiphytical orchid, spread throughout Java between 300 and 1500 m. above sea-level, in forests and on trees in native villages. The reproduction is effected by seeds.

Collected, according to the Dutch text, at an altitude of 700 m. The English text gives an altitude of 208 feet i.e. between 60 and 65 m.

39 (31). *Liparis* spec.

Collected at an altitude of ± 50 m. Epiphytic.

40 (37). *Thrixspermum* spec.

Genus of epiphytical orchids, in the plains and the lower mountainous regions of Java represented by several species. Some species are found in rather densely shadowed localities, others in sunny or but slightly shadowed ones. The reproduction is effected by seeds.

A species was collected at an altitude of 700 m.

Piperaceae.

41 (20). *Heckeria umbellata Kth.* var. *subpeltata D.C.* (= H. subpeltata Willd.). See p. 191, No. 17.

Collected at an altitude of ± 100 m.

Ulmaceae.

42 (25). *Trema orientale Bl.* (= Tr. amboinense Bl.). See p. 132, No. 15 and § 5 herebefore.

As it appears from § 5, collected in the lower ravines on the south-east- or east-side of the island.

Moraceae.

43 (10). *Ficus ampelas L.*

Small tree, spread throughout Java from the plains up to an altitude of 1200 m., in thickets, in teak-forests and secondary and (less often) primary rain-forests, in and around native villages and sometimes in unkept gardens. The reproduction is effected by the small figs which are copiously produced.

Collected at an altitude of 100 m.

44 (11). *Ficus fistulosa Reinw.* See p. 133, No. 16 and §§ 5 and 6 herebefore.

Collected on the summit of Mount Rakata.

45 (12). *Ficus fulva Reinw.* See p. 133, No. 17 and § 5 herebefore.

Collected „in the lower part of the isle" (D. v. L.).

46 (13). *Ficus lepicarpa Bl.*

Small tree spread throughout Java between 400 and 1500 m. above sea-level in secondary and primary rain-forests and in other shadowed moist localities, especially on or near water-sides. The reproduction is effected by the figs.

The English text gives no habitat. The Dutch text says: „in the lower parts".

47 (15). *Ficus quercifolia Roxb.* See p. 192, No. 20.

According to the Dutch text collected at altitudes of 100 and of 800 m. The Englisch text says: „*No real forestplant*", but gives no altitude.

48 (16). *Ficus retusa L.* See § 6 herebefore.

Large tree, spread throughout Java from the plains to an altitude of 1200 m., especially in light forests, on the inner borders of mangrove forests and in not too dry parts of teak-forests. It is frequently cultivated as an ornamental tree. The reproduction is effected by the figs which are produced in enormous numbers.

According to the Dutch text collected at an altitude of 700 m. The English text says: „in the higher parts of Mount „Rakata".

49 (17). *Ficus ribes Reinw.* See § 6 herebefore.

Collected „in the higher parts of Mount Rakata". The Dutch text adds: „up to the summit". This is in accordance with § 6. But in a subsequent paper [1]) an altitude of 300 m. (consequently the lower half of the mountain) is given for specimens of this species found in April 1919.

50 (14). *Ficus septica Burm.* (= F. leucantatoma Poir.).

Shrub, spread throughout Java from the plains up to an altitude of 1200 m., on road-sides, in thickets and second-growth forests, in teak-forests, in and around native villages and very often in unkept gardens. The reproduction is effected by figs which are greedily eaten by bats.

Collected — as appears from the Dutch text — in the lower parts of the island. The English text only says: *„No real „forest-plant";* the Dutch text adds: *„as a rule no forest-plant".*

51 (18). *Ficus toxicaria L. fil.* See p. 115, No. 44.

Collected in the lower ravines.

52 (19). *Ficus variegata Bl.* See herebefore § 5 where this species is mentioned as „one of the many trees belonging to a „few species" that were growing in a dry water-channel.

Collected at an altitude of 100 m. In a paper to be discussed later on [2]) Mr. Docters van Leeuwen asserts that this species occurred only sporadically on Krakatao.

---

[1]) Dr. W. Docters van Leeuwen, The Galls of the Islands of the Krakatau-group and of the Island of Sebesy, in Bull. Jard. Bot. Buitenzorg, Serie III, Vol IV (1922), p. 298.

[2]) Dr. W. Docters van Leeuwen, The Vegetation of the Island of Sebesy, situated in the Sunda-Strait, near the Islands of the Krakatau-group; in the year 1921, in Ann. Jard. Bot. Buitenzorg XXXII (1923), p. 161.

Urticaceae.

53 (27). *Villebrunea rubescens Bl.* See § 5 and § 6 herebefore.
According to Appendix II collected at an altitude of 200 m. But in § 5 this species is mentioned as occurring in the dry water-channel beneath an altitude of 100 m. (at which height the way was blocked by an unscalable precipice). In 1922 found [1]) at altitudes of 150 and 700 m.

54 (26). *Leucosyke capitellata Wedd.* See p. 192, No. 23.
Collected in the lower ravines.

Rutaceae.

55 (22). *Citrus* spec.
Several species of this genus are extensively cultivated in Java; none are wild, though in favorable localities sometimes a young tree may arise from a pip thrown away by a person eating the fruit. But in rain-forests such young trees are exceedingly rare.
Collected near the summit of the mountain.

Euphorbiaceae.

56 (5). *Macaranga tanarius M. A.* See p. 135, No. 26 and § 5 herebefore.
Collected from the sea-shore up to an altitude of 100 m.

57 (4). *Homalanthus populnea* O. *K.* See p. 193, No. 25 and § 5 herebefore.
Collected in ravines between 30 and 200 m. above sea-level. But in a subsequent paper [2]) an altitude of ± 300 m. is given for this species at this same trip.

Vitaceae.

58 (28). *Leea aequata L.* See p. 136, No. 29.
The English text mentions no habitat but says only: „*No „real forest-plant*". The Dutch text adds: „close to the coast".

Sterculiaceae.

59 (24). *Melochia umbellata Stapf.* See p. 117, No. 51 and § 5 herebefore.
Collected in the lower ravines.

---

[1]) Dr. W. Docters van Leeuwen, The Galls of the Islands of the Krakatau-group and of the Island of Sebesy in Bull. Jard. Bot. Buitenz., Série III, Vol. IV (1922), p. 310.

[2]) Dr. W. Docters van Leeuwen, The Galls of the Islands of the Krakatau-group and of the Island of Sebesy in Bull. Jard. Bot. Buitenz., Série III, Vol. IV (1922), p. 301.

Melastomataceae.

60 (9). *Melastoma malabathricum L.* See p. 150, No. 33.

According to the English text: „prevailing on naked moun-„tain-slopes, *no real forest-plant*"; according to the Dutch text: „mostly on slopes bearing but a scanty vegetation, no „forest-plant". It is hardly to be believed that this *Melastoma* was really a prevailing, i. e. a *predominant* and not a scattered plant on *naked* slopes. Perhaps the translator has not rightly understood the Dutch text.

Solanaceae.

61 (23). *Solanum blumei Nees.* Determination doubtful.

Rather robust herb, spread throughout the moister parts of West-Java in rain-forests between 1000 and 2500 m. above sea-level, usually in scattered specimens. The reproduction is effected by seeds.

According to the Dutch text collected at an altitude of 700 m., according to the English text at an altitude of 2000 feet ($\pm$ 600 m).

Bignoniaceae.

62 (1). *Radermachera glandulosa Miq.* See § 5 herebefore.

According to the Dutch text „scattered in the ravines „between 30 and 500 m. above sea-level. Considerable-sized „trees". According to the English text occurring between 150 and 1500 feet, i. e. between 45 and 450 m.

Gesneriaceae.

63 (7). *Trichosporum pulchrum Bl.*

Epiphyte or less often a chasmophyte, spread throughout the moister regions of Java between 50 and 1750 m. above sea-level. It grows in shadowed localities, mostly in forests either primary or secondary, not rarely also on trees planted in native villages or along road-sides, less often on rocks. The reproduction is effected by the small seeds, which bear on both ends a long hair serving as a kind of sail to the seed, which is easily carried along by the wind.

Found at an altitude of 400 m. Epiphytic.

64 (8). *Trichosporum volubile Nees.*

Creeping or climbing plant, usually found in very wet localities [1]), occurring in West-Java from the plains up to an altitude of ± 300 m. The reproduction is effected by seeds, which in the same way as those of the preceding species are provided with 2 hairs.

Collected on the summit of Mount Rakata. The Dutch text adds: „in brushwood" but in § 6 herebefore it is said that the plant crept over *Saccharum spontaneum* and 2 herbaceous fern-species.

65 (6). *Cyrtandra sulcata Bl.* See p. 194, No. 29 and § § 6 and 15. From an altitude of 50 m. to the top of the mountain.

Rubiaceae.

66 (21). *Nauclea purpurascens Bl.* See p. 194, No. 30 and § 5 herebefore.

From an altitude of 30 m. to the top of the mountain. The Dutch text says from 30 m. to *near* the top.

Compositae.

67 (2). *Erechthites valerianifolia Rafin.*

Annual weed of cultivated fields, in 1845 accidentally introduced from Brazil into Java, at present spread throughout the moister regions of the island from the plains up to an altitude of 2200 m., especially above 400 m. in tea- and Cinchona-plantations, in fields, on road-sides and in recent clearings, locally (especially on fertile soils) often a very common plant. The reproduction is effected by the fruits which are provided with a well-developed pappus.

Collected near the summit of the mountain. „*No real „forest-plant*" (D.v.L.) The Dutch text adds: „found between „shrubs".

68 (3). *Gynura sarmentosa D.C.*

Straggling or scandent weed, often with long depending branches, spread throughout Java from the plains up to an altitude of 1600 m., but only in regions where the east-monsoon is not very strong. It grows in thickets and hedges,

1) Edeling (Natuurkundig Tijdschrift Ned.-Indië XXXI (1870), p. 324) found it in a bog near Meester Cornelis (south of Batavia); I myself saw it climbing in a small *Metroxylon*-morass south of Buitenzorg.

along forest-paths, on woodborders, in light woods. The reproduction is effected by the fruits, which are provided with a well-developed pappus.

Collected at an altitude of 600 m.

In my discussion of § 14 herebefore I mentioned already that Mr. Docters van Leeuwen has singled out ten plants from the list of „forest-plants" above (No. 9, 12, 14, 25, 31, 47, 50, 58, 60 and 67) as being „no real forest-plants". Consequently there remain 58 species which he considers to be „real forest-plants". One is at some pains to ascertain what Mr. Docters van Leeuwen may have understood by this term. He can hardly mean plants which sometimes may be found in any kind of forest, for if so, there would be but comparatively few species on Krakatao that were no forest-plants. We have therefore to suppose that his intention was to restrict this term to plants which, as a rule, are found in forests and are comparatively rare out of them. There are in Java several kinds of forests, each with their characteristic plants, but the word forest-plant cannot have been taken here in so wide a sense that it comprises all these plants. Mr. Docters van Leeuwen cannot have meant plants of the coastal forests, for none of the littoral plants found on Krakatao, not even *Casuarina equisetifolia L.*, which formed a small forest near the sea, are included in the list of forest-plants. Teak-forests are on Krakatao out of the question and so are true savanna-forests, though young secondary forests, gradually replacing grassy wilds, may sometimes resemble them. Neither can young secondary rain-forest be meant, for in such forests No. 14, *Drymoglossum heterophyllum C. Chr.*;– No. 25, *Lygodium scandens Sw.*;– No. 50, *Ficus septica Burm.*;– No. 58, *Leea aequata L.* and No. 60, *Melastoma malabathricum L.*, considered by him to be *no real forest-plants*, are plants of common occurrence. Consequently Mr. Docters van Leeuwen can by „real forest-plants" only mean plants of primeval and rather old secondary rain-forests. At present Krakatao does not bear any very old forests, the former woods having been devastated by the eruption of 1883 (though locally plants, rhizomes, seeds or spores quite well may have survived). But in many ravines of the island secondary forests have grown up since the eruption, in the higher parts of the cone (barring those quite near the top) already rather advanced in age, dark and humid; the lower ones gradually younger, thinner, consequently sunnier and drier. Considering that the forests in the higher parts of the mountain differ from those beneath in age, density and humidity, taking into account that the walls of the ravines in which these forests are growing are often almost perpendicular and fit only for a kremnophytic vegetation which has nothing to do with forest but may develop on any steep not too dry neither too sunny bank and not forgetting that Krakatao has been frequently visited by man and has been inhabited during more than two years by many people (which fact must have influenced the vegetation) we can expect beforehand that the *real forestplants* of Mr. Docters van Leeuwen will prove to belong to very divers oecological groups. This preopinion is fully confirmed by the list given in Appendix III, which, even after those *„forest-plants which are no real forest-plants"* have been excluded, still consists of very heterogeneous elements many of which are no forest-plants at all.

To begin with, No. 55, *Citrus* spec., is certainly no forest-plant. In Java there occur no species of this genus in a truly spontaneous state. – No. 28. *Equisetum debile Roxb.*, is likewise no forest-plant at all, much less a „real forest-plant". Mr. Docters van Leeuwen himself found it on a slope bearing but a meagre vegetation. The plant requires much light and though sometimes occurring on the borders of light woods and in glades never penetrates into the forest proper. – No. 7, *Nephrolepis exaltata Schott* – and especially the form recorded by Mr. Docters van Leeuwen as *N. hirsutula Presl*, – is a fern of grassy wilds and low sunny thickets. Mr. Docters van Leeuwen himself admits that he found this „forest-plant" mostly in little shadowed localities. – No. 27, *Ophioglossum reticulatum L.*, a plant of badly pervious soils is rather frequently found in teak-forests, but much more often on road-sides, on premises and lawns, on the small grass-clad dikelets (galangans) between rice-fields and often also above such nests of white ants, as are covered on their upper side with an impervious layer of clay. It has no right at all to be called a „real" forest-plant. – No. 4, *Dryopteris setigera O.K.* is a kremnophyte of sunny or moderately shadowed not too dry localities; it is often found on almost perpendicular water- and road-sides and is certainly no real forest-plant. – No. 5, *Aspidium melanocaulon Bl.*, likewise a kremnophyte, but of more humid and well shadowed localities is often found on steep sides of water-courses and in native villages on the margins of wells. Though it may be found in forests it is certainly no typical, no real forest-plant. – No. 24, *Lygodium circinatum Sw.*, a plant of moderately shadowed not too dry localities, may be found in secondary forests and on forest-borders but is much more frequent in thickets and hedges. — No. 35, *Eulophia macrostachya Lindl.* is an orchid of bamboo-plantations, of thickets and young secondary forests, but no „real" forest-plant. — No. 41, *Heckeria umbellata Kth.* var. *subpeltata Willd.* grows in shadowed, constantly humid localities, on ravine-sides, along water-courses, in thickets, much more frequently out of forests than within them. — No. 64, *Trichosporum volubile Nees*, a plant of humid localities, often creeping over shrubs or grasses has hardly a right to be called a forest-plant and is certainly no „real" one. — No. 23, *Schizaea dichotoma Sm.*, in Java seemingly a rare species, may be found in rain-forests on not too dark and not too humid ridges, but it occurs more often in bamboo-plantations and in teak-forests. — No. 30, *Selaginella plana Hieron.*, a plant of shadowed localities, is frequently found in young secondary forests but at least as often in thickets and on humid river-banks. Neither of these two species are „real" rain-forest-plants., — No. 44, *Ficus fistulosa Reinw.*, — No. 45, *Ficus fulva Reinw.*, — No. 51, *Ficus toxicaria L.* and No. 57, *Homalanthus populnea O.K.* are trees of grassy jungles, of ravine-sides, glades, forest-borders and young secondary forests; they need much light and can hardly be called real forest-plants. Neither is No. 48, *Ficus retusa* L., a real rain-forest-plant; it is often found growing as an epiphyte on trees in native villages and even on houses; it occurs also on the inner border of mangrove-forests and in teak-forests. — No. 56, *Macaranga tanarius M. A.* and No. 59, *Melochia umbellata Stapf* are frequently found in thickets and in deserted fields, which are being gradually invaded by jungle; they may occur also in teak-forests but are no real forest-plants. They have no more right to be taken for such than *Pipturus incanus Wedd.*, and *Melastoma malabathricum L.* which were quite rightly excluded by Mr. Docters van Leeuwen from his list of forest-plants. No. 15, *Hymenolepis brachystachys J. Sm.*, — No. 16, *Hymenolepis spicata Presl*, — No. 19, *Polypodium phymatodes L.*, — No. 20, *Polypodium punctatum Sw.* and. No. 22, *Cyclophorus adnascens Desv.*, may be found in rain-forests but in not too dry regions they are very frequent also on trees planted along road-sides, in gardens, in native villages and church-yards. They are by no means real forest-plants. Neither is No. 34, *Nervilia aragoana Gaud.*, a real rain-forest-plant. It is as a rule an inhabitant of jungles, of bamboo-woods and of teak-forests and is but rarely found out of them.

Nothing can be said of No. 8, *Davallia* spec.,— No. 21, *Polypodium* spec.,— No. 32, *Habenaria* spec.,— No. 37, *Oberonia* spec.,— No. 39, *Liparis* spec. and No. 40, *Thrixspermum* spec., of which plants only the generic names are given. Species of these genera being found as well out of forests as within them, it is impossible to decide whether they are „real" forest-plants.

Consequently of the 58 „real" forest-plants of Mr. Docters van Leeuwen there remain at best 26 which as a rule occur in rain-forests of some age. As to their habitat I refer to the list given above. Those plants are: No. 1, *Trichomanes humile Forst.*,— No. 2, *Alsophila* spec.,— No. 3, *Dryopteris megaphylla C. Chr.*,— No. 10, *Diplazium polypodioides Bl.* (but also rather frequent in shadowed, moist localities, out of forests),— No. 11, *Asplenium nidus L.* (frequently found out of forests),— No. 13, *Antrophyum* spec. (all Javanese species of this genus),— No. 17, *Polypodium accedens Bl.*,— No. 18, *Polypodium heracleum Knze.*,— No. 26, *Angiopteris evecta Hoffm.*,— No. 29, *Lycopodium squarrosum Forst.*,— No. 33, *Peristylus gracilis Bl.*,— No. 36, *Oberonia monstruosa Lindl.*,— No. 38, *Liparis viridiflora Lindl.*,— No. 42, *Trema orientale Bl.* (very often also in deserted fields),— No. 43, *Ficus ampelas L.* (frequently found in gardens and in native villages),— No. 46, *Ficus lepicarpa Bl.*,— No. 49, *Ficus ribes Bl.*,— No. 52, *Ficus variegata Bl.* (frequently fount out of forests),— No. 53, *Villebrunea rubescens Bl.*,— No. 54, *Leucosyke capitellata Wedd.*,— No. 61, *Solanum blumei Nees* (provided the determination be correct),— No. 62, *Radermachera glandulosa Miq.*,— No. 63, *Trichosporum pulchrum Bl.* (in moist regions also frequent on trees in native villages),— No. 65, *Cyrtandra sulcata Bl.*,— No. 66, *Nauclea purpurascens Bl.*,— No. 68, *Gynura sarmentosa D.C.* (very often also out of forests.).

Besides the five littoral species excluded without reason from Appendix II and recorded already by me on p. 247, Mr. Docters van Leeuwen mentions in Appendix I to his paper 11 or 12 other species as found in 1919 on Krakatao which are not given in Appendix II and III and therefore probably were neither found on or near the beach nor were forest-plants, not even unreal ones. These species, about the habitat of most of which we are left quite in the dark, are:

Polypodiaceae.

1. *Dryopteris unita O.K.* See p. 56, sub No. 7.

According to § 6, herebefore collected on the higher ridges and on the top the mountain.

2. *Dryopteris flaccida O.K.* See p. 57, No. 8.

This species, an inhabitant of humid dark forests is recorded by Mr. Docters van Leeuwen only for the sixth trip of April 1919. A habitat is not given. If it was really found on this excursion it is hardly to be believed that it was not collected in the forest. But as it is not given in the list of forest-plants it may have crept into Appendix I by some error. Treub wrongly mentioned this species for the first trip in 1886; the plant in question was *Dryopteris setigera*. In 1919/1920 when Mr. Docters van Leeuwen published his papers, this error had not yet been detected, yet he does not mention the species for Treub's excursion. Therefore in Appendix I the figure 6 may be only a misprint for 1.

3. *Aspidium* spec.

All Javanese species of *Aspidium* are inhabitants of well-shaded and rather humid localities.

Habitat not mentioned.

4. *Polypodium longissimum Bl.* (Pleopeltis longissima Moore).

Terrestrial fern, spread throughout the moister regions of West- and Central-Java from the plains up to an altitude of 900 m. It grows on water-sides, in hedges, in grassy wilds and jungles, in bamboo-plantations.

Habitat not mentioned.

Lycopodiaceae.

5. *Lycopodium cernuum L.* See p. 113, No. 29.

This species – like *L. squarrosum Forst.* – is for the sixth trip (April 1919) given in Appendix I to the Dutch text but omitted in Appendix I to the English text, giving all plants found on Krakatao.

Habitat not mentioned.

Orchidaceae.

6. *Arundina speciosa Bl.* See p. 107 and §§ 5 and 16.

In § 5 herebefore Mr. Docters van Leeuwen mentions that this species, „abundant at the time of former explorations „had (in April 1919) been ousted by the spreading forest". Nevertheless it is recorded in Appendix I as found in April 1919.

Habitat not mentioned.

Urticaceae.

7. *Pouzolzia zeylanica Benn.* See p. 181, No. 20.
Habitat not mentioned.

Euphorbiaceae.

8. *Acalypha indica L.*
Annual ruderal weed, spread throughout the lower regions of Java from the plains to an altitude, as far as is known at present, of 400 m. It occurs by preference in somewhat shaded localities, on road-sides, in gardens, fields, lawns and church-yards, in native villages and in teak-forests. It is exclusively reproduced by seeds which are produced in great numbers and are flung away when the fruits burst.
Habitat not mentioned.

9. *Ricinus communis L.*
Shrub or very small tree, very often cultivated in gardens and in fields, especially by the natives, and in waste places frequently occurring in a semi-spontaneous state. The reproduction is effected by the fruits, which are produced in great numbers. Each of them contains 3 seeds, which are flung away the moment the fruit bursts.
Habitat not mentioned.

Solanaceae.

10. *Capsicum frutescens L.*
Annual, in Java from the plains up to an altitude of 1000 m. very often cultivated for its fruits and near human habitations frequently found in a spontaneous or subspontaneous state. The reproduction is effected by seeds which are produced in great numbers.
Habitat not mentioned.

Scrophulariaceae.

11. *Stemodia verticillata Bold.* (= St. parviflora Ait.).
Small annual herb of American origin, introduced into Java already long ago and at present rather common in the western half of the island form the plains up to 1000 m. above sea-level. It occurs in fields and gardens, often also on gravelled paths. It is reproduced by its seeds, which are easily spread by floating rain-water.
Habitat not mentioned.

Cucurbitaceae.

11. Luffa cylindrica Roem. See p. 187, footnote 1.
Habitat not mentioned.

Sixty-six species of Phanerogams and higher Cryptogams collected in 1919 on Krakatao were then for the first time found there, viz: *1, Acalypha indica L.;* – *2* (III, 26)[1]), *Angiopteris evecta Hoffm.;* – *3* (III, 13), *Antrophyum* spec.; – *4* (II, 30), *Aristolochia tagala Cham.;* – *5* (III, 5), *Aspidium melanocaulon Bl.;* – 6 *Aspidium* spec.; – *7* (III, 11), *Asplenium nidus L.,* – *8, Capsicum frutescens L.;* – *9* (II, 44), *Carica papaya L.;* – *10* (III, 55), *Citrus* spec.; – *11* (II, 31), *Cyathula prostrata Bl.;* – *12* (II, 7), *Cyclophorus acrostichoides Presl;* – *13* (III, 22), *Cyclophorus adnascens Desv.;* – *14,* (II, 20), *Cyperus cylindrostachyus Boeck.;* – *15* (III, 8), *Davallia* spec.; – 16 (II, 26), *Dioscorea spec.;* – *17,* (III, 10), *Diplazium polypodioides Bl.;* – *18* (III, 14), *Drymoglossum heterophyllum C. Chr.;* – *19,* (III, 3), *Dryopteris megaphylla C. Chr.;* – *20* (II, 19), *Eleusine indica Gärtn.;* – *21* (II, 64), *Eleutheranthera ruderalis Sch. Bip.;* – *22* (III, 28), *Equisetum debile Roxb.;* – *23* (III, 67), *Erechthites valerianifolia Rafin.;* – *24* (III, 16), *Eriochloa ramosa O.K.* (or perhaps *E. subglabra Hitchc.*); – *25* (III, 35), *Eulophia macrostachya Lindl.;* – *26* (III, 42), *Ficus ampelas L.;* – *27* (III. 46), *Ficus lepicarpa Bl.;* – *28* (III, 48), *Ficus retusa L.;* – *29* (III, 49), *Ficus ribes Reinw.;* – *30* (III, 50), *Ficus septica Burm.;* – *31* (III, 52), *Ficus variegata Bl.* – *32* (III, 68), *Gynura sarmentosa D. C.;* – *33* (III, 32), *Habenaria* spec; – *34* (II, 17), *Hoplismenus compositus P. B.;* – *35* (III, 15), *Hymenolepis brachystachys J. Sm.:* – *36* (III, 16), *Hymenolepis spicata Presl;* – 37 (II, 22), *Kyllinga monocephala Rottb.;* – 38 (II, 52), *Lantana camara L.;* – 39 (III, 38), *Liparis viridiflora Lindl.;* – 40 (III, 39), *Liparis* spec.; – 41 (III, 29), *Lycopodium squarrosum Forst.;* – 42 (III, 24), *Lygodium circinatum Sw.;* – 43 (III, 25), *Lygodium scandens Sw.;* – 44, *Mucuna gigantea D.C.;* – 45 (II, 2; III, 6), *Nephrolepis biserrata Schott;* – 46 (III, 36), *Oberonia monstruosa Lindl.;* – 47 (III, 37), *Oberonia* spec.; – 48 (III, 11), *Odontosoria chinensis J. Sm.;* – 49 (II, 55), *Oldenlandia subulata Korth. (?);* – 50 (III, 33), *Peristylus gracilis Bl.;* – 51 (III, 17), *Polypodium accedens Bl.;* – 52 (III, 18), *Polypodium heracleum Knze.;* – 53, *Polypodium longissimum Bl.;* – 54 (III, 21), *Polypodium* spec.; – 55 (III, 62), *Radermachera glandulosa Miq.;* – 56, *Ricinus communis L.;* – 57 (III, 23), *Schizaea dichotoma Sw.;* – 58, (III, 61), *Solanum blumei Nees.;* – 59, *Stemodia verticillata Bold.;* – 60 (II, 65), *Synedrella nodiflora Gärtn.;* – 61 (III, 40), *Thrixspermum* spec.; – 62 (II, 32), *Tinospora coriacea Beumée;* – 63 (III, 1), *Trichomanes humile*

[1]) The numbers within brackets refer to my rearrangement of Appendix II (See pp. 234 seq.) and III (See pp. 249 seq.) of Mr. Docters van Leeuwen's paper.

Forst.; — 64 (III, 63), *Trichosporum pulchrum Bl.;* — 65 (III, 64), *Trichosporum volubile Nees;* — 66 (III, 53), *Villebrunea rubescens Bl.* — As I have already repeatedly set forth we may not conclude from a list of finds in a comparatively small part of Krakatao that such plants as were not collected at former excursions, likewise made in restricted localities, were not yet present then on the island. This is the more obvious if we bear in mind that in 1919 the top of the island (somewhat upwards of 800 m.) was reached, whilst in 1908 no greater height was attained than 400 m. At both excursions only small parts of the island were visited and even those small parts were but very cursorily investigated. Such flying investigations do not allow of drawing any conclusions as to the absence of a species, which may have been overlooked or not found because the locality it was growing in remained unexplored. Hence it is impossible to ascertain which of the above mentioned 66 species were introduced after 1908, if such might have happened. Anyhow, such an introduction has to be considered probable. Bearing in mind that Mr. Händl and his household have lived during 2 years on Krakatao, that he has laid out a garden there and has introduced domestic animals (swine, geese, hens) that the laying out of a garden and the keeping of animals always cause ruderal weeds to be introduced, we may with some right suppose that such plants of cultivation and weeds of gardens, fields, road-sides and waste places as were for the first time found in 1919 *may* have been purposely or accidentally introduced by Mr. Händl, his followers and his livestock. These are 13 in number, viz. *Acalypha indica L.* — *Capsicum frutescens L.* — *Cyathula prostrata Bl.,* — *Cyperus cylindrostachyus Boeck.,* — *Eleusine indica Gärtn.,* — *Eleutheranthera ruderalis Sch. Bip.,* — *Erechthites valerianifolia Rafin.,* — *Eriochloa ramosa O.K.* (or perhaps *E. subglabra Hitchc.*), — *Hoplismenus compositus P. B.,* — *Kyllinga monocephala Rottb.,* — *Ricinus communis L.* — *Stemodia verticillata Bold.* and *Synedrella nodiflora Gärtn.* But even for this introduction there exists no conclusive evidence, except as regards *Capsicum frutescens L.* (See p. 197).

One of the 54 remaining species, *Carica papaya L,* had been cultivated by Mr. Händl during his two years' stay on the island. As the coast was frequently visited by Mr. Händl's people for the purpose of gathering wood, the specimens found by Mr. Docters van Leeuwen *may* have been the offspring of these cultivated specimens. But such need not at all have been the case, as in 1906 and 1908, long before Mr. Händl established himself on Krakatao, the tree was already growing on Lang Eiland and Verlaten Eiland. Any visitors of Krakatao, native fishermen as well as Europeans, may have introduced it, neither is introduction by animals to be considered impossible. *Nothing definite is known.* — *Lantana camara L.* had in 1908 already been found on Verlaten Eiland near the coast. It may have been introduced into Krakatao from there or from any other of the surrounding islands, either by man, by animals or perhaps by the sea. Though the kernels of the fruit do not possess floating power, fruit-bearing branches, carried by swollen rivers to the sea, might have been washed ashore. *Here again nobody can tell what has happened.* — Of the species of *Dioscorea* the identity is not fully established. Several species of this genus are cultivated in Java by the natives for their tubers which may reach an enormous size and form a not unimportant article of food. In cultivation the plant is always reproduced by its tubers. As it is not even known which species was found, it is quite useless to speculate on possible modes of introduction. Of the remaining species there are 25 ferns and fern-allies, viz. *Angiopteris evecta Hoffm.*, — *Antrophyum* spec., — *Aspidium melanocaulon Bl.*, — *Aspidium* spec., — *Asplenium nidus L.*, — *Cyclophorus acrostichoides Presl*, — *Cyclophorus adnascens Presl*, — *Davallia* spec., — *Diplazium polypodioides Bl.*, — *Drymoglossum heterophyllum C. Chr.*, — *Dryopteris megaphylla C. Chr.*, — *Equisetum debile Roxb.*, — *Hymenolepis brachystachys J. Sm.*, — *Hymenolepis spicata Presl*, — *Lycopodium squarrosum Forst.*, — *Lygodium circinatum Sw.*, — *Lygodium scandens Sw.*, — *Nephrolepis biserrata Schott*, — *Odontosoria chinensis J. Sm.*, — *Polypodium accedens Bl.*, — *Polypodium heracleum Knze.*, — *Polypodium longissimum Bl.*, — *Polypodium* spec., — *Schizaea dichotoma Sw.*, and *Trichomanes humile Forst.* which may be reproduced by their very light spores. These may quite well have been introduced by wind, but this is not proven; other means of introduction are possible. It is not even proven that all of them were introduced after 1883. The earth-ferns, and perhaps also some of the epiphytes, may very well be the offspring of plants that had escaped destruction. *Nobody knows.* One can only guess and authors have repeatedly made hazardous guesses. So has Mr. Docters van Leeuwen done, as he admits himself in the Dutch text of his paper [1], in Appendix I,

[1] Dr. W. Docters van Leeuwen in Handelingen Eerste Ned. Ind. Natuurwetensch. Congres, 1920, p. 51: „ik heb bij benadering de verspreidingswijze aangegeven, maar ik heb er dikwijls een slag naar moeten slaan". The English rendering of this text in Ann. Jard. Bot. Buitenzorg XXXI (1921) p. 122: „I have stated the manner of spreading „with the nearest approach to accuracy I could, but I have often been compelled to „guess", is quite inaccurate. The correct translation runs as follows: „I have roughly stated „the manner of spreading, but I was often compelled to make quite hazardous guesses".

giving a rather inaccurate list of all plants found till 1919 on the 3 islands of the Krakatao group, where he has added to the name of each plant its means of distribution which, as appears from § 17 herebefore, is considered by him also as the means of introduction. His assertions in this matter are not based on actual observation but, as I have already pointed out in the introductive chapter, he was, like former authors on the subject, only guided by the quite unwarranted opinion that Cryptogams and such Phanerogams as possess light or winged seeds were carried over by wind, plants with fleshy, succulent or adhering fruits by birds or bats, the rest by the sea. Sometimes, without giving however any reason for his opinion, he adscribes introduction to man. It needs scarcely saying that guess-work like this is entirely devoid of value and cannot be seriously discussed. I can only state that *nobody knows* which plants of the interior have been introduced into Krakatao after the eruption and that *nothing at all* is known of the mode of introduction.

Appendix IV to the Dutch text of the paper of Mr. Docters van Leeuwen contains a list of galls found on Krakatao. As this Appendix is not found in the English text and the subject is more extensively treated by the same author in another paper [1]), I shall revert to it in my discussion of that publication (Chapter XII). Appendix V to the Dutch text, likewise omitted in the English text, gives a list af all birds seen by Mr. Bartels on Krakatao and Verlaten Eiland from April 24–29th 1919. For the study of the revegetation of the island it has at present no importance.

If we now proceed to sum up the real results of the trips of 1919 we can only state that in the years elapsed since 1908 no new elements were found to have appeared in the littoral flora. The secondary forest of the interior had probably – as might be expected beforehand – extended considerably downwards. The lower parts of it had presumably become denser, richer in trees and in undergrowth. Whether these additions had come over from such parts of the forest as were not examined before 1919 or whether they had been introduced since 1908 from outside the island *is absolutely unknown*. Rather many epiphytes were found, a phenomenon of common occurrence in old secondary forests in wet climates. These epiphytes, mostly ferns and orchids, *may* have been introduced after the eruption, the epiphytic plants in all likelihood having been quite

1) Mr. W. Docters van Leeuwen, The Galls of Krakatao and Verlaten Eiland in 1919, in Ann. Jard. Bot. Buitenzorg XXXI (1920), p. 57 seq.

destroyed in 1883, though spores and seeds may have survived. As to the mode of introduction of these epiphytes nothing is known with certainty. It may have been done by the wind. It is probable, though not proven, that after 1908 a number of ruderal weeds were introduced by human influence. – Very many fungi were found, especially in the moist ravines. Their spores *may* have been introduced by the wind after 1883 though such need not at all to have happened. Other modes of introduction are possible, a number of mycelia may have survived the eruption.

So the expeditions of 1919 have only confirmed the well-known rule that a secondary tropical forest protected from, or at least not continually or too heavily attacked by its three most dangerous foes, man, cattle and fire, tends to increase in size and in the number of species. But these expeditions have contributed nothing at all to answering the questions put on the first page of this paper. By the answers to these questions we should measure the success of the endeavours to solve the Krakatao-problem. Hence the botanical expeditions to Krakatao in 1919, when considered from a scientific view-point, were a complete failure.

XII. FURTHER PUBLICATIONS ON THE REVEGETATION OF KRAKATAO.

A. – Dr. W. Docters van Leeuwen, The Galls of Krakatau and Verlaten Eiland (Desert Island) in 1919, in Ann. Jard. Bot. Buitenzorg XXXI (1921), p. 57 seq.

This paper very shortly describes 24 galls found on 19 species of plants. Of these 24 galls 13 are caused by mites, 7 by gnats [1]), 3 by Psyllidae and one by a species of Aphis. The animals acting as agents are not described at all and not any light is shed on the manner in which they came over to Krakatao; only vague possibilities are discussed. Mr. Docters van Leeuwen cites as a remarkable fact [2]) that a Psyllida-gall was found on *Thespesia lampas*. This is certainly remarkable, for *Thespesia lampas*, a shrub of teak-forests and grassy jungles in regions with a well-marked east-monsoon, was never yet found on any of the islands of the Krakatao-group, which form a very unfit habitat for it. The same author asserts that „he did find it in a „table of plants [3]) found by Dr. Valeton in 1904". But this is hardly possible, as Dr. Valeton visited Krakatao not in 1904 but in March 1905 (see Chapter VII) and the only species of *Thespesia* mentioned by him is *Th. populnea Soland* [4]), a sea-side tree of which he found young specimens. Dr. Docters van Leeuwen states that he found *Th. lampas* on Verlaten Eiland but, as it appears from the lists published elsewhere by him [5]), it was not this species he found but *Th. populnea*.

Mr. Docters van Leeuwen states [6]): „most of the galls „I found on genuine coast-plants or on plants occurring within „a short distance of the shore. There it was also that the veget- „ation comprised the greatest variety of species. On climbing „the mountain-slopes the number of different phanerogamous „plants becomes gradually smaller whilst, contrarily, the number of „individuals belonging to the same species increases considerably".

---

1) In a subsequent paper (Vide sub B) called midges.
2) l. c. p. 63.
3) See p. 129, footnote 2.
4) See p. 137, No. 31.
5) Ann. Jard. Bot. Buitenzorg XXXI (1921), pp. 79, 128, 137.
6) Ann. Jard. Bot. Buitenzorg XXXI (1921), p. 63.

This statement I think is not sufficiently warranted. In Appendix II to his paper on the excursions of 1919 (See p. 234 seq.), Mr. Docters van Leeuwen records 58 Phanerogams as found on or near the beach of Krakatao to which number should be added 5 species (See p. 247), that were certainly found on or near the beach though they were not recorded in Appendix II, hence in total 63 species. In Appendix III he mentions 38 Phanerogams considered by him to be forest-plants, hence found in the interior. Large as the difference between the number of plants found on or near the shore and that in the interior may be, we have no right to conclude from it that the shore is richer in Phanerogamic species than the interior. There is another conclusion which seems to me far more correct. The beach, the sandy parts of it at least, can be very comfortably investigated throughout a length of more than 4 km. [1]); most of the species growing on it are herbaceous or shrubby and, being well exposed to the light, flower freely, hence are apt to be collected. The interior offers much more difficulties to the investigator and many species there are either arboreous or, by cause of want of light, do not flower in which case they are easily overlooked. Experience teaches that unsufficiently trained collectors in the tropics often neglect to gather materials from arboreous or non-flowering species. *Hence they easily get the impression that a forest is much poorer in species than it really* is. Moreover, but a *very small part* of the innumerable ravines and ridges of Krakatao can have been examined in 1919 and one often sees that ravines and ridges but little remote from each other differ in vegetation. By examining but a small part of the ravines and ridges one gets but a part of the species present. In 1923, 4 years after the trips of 1919, Mr. Docters van Leeuwen asserted [2]) that a great many *new* plants, especially *Fungi, Musci, Hepaticae* and *Lichenes,* had been noticed on Krakatao. It is much less probable that these species had newly settled there in those 4 years than that many plants present already in 1919 were overlooked in that year. From the results published it does not at all appear that the investigated *lower* slopes, covered with a dense *Saccharum*-jungle intermixed with scattered

1) See p. 64, footnote 1.
2) Ann. Jard. Bot. Buitenzorg XXXII (1923), p. 157.

trees and small groves, were in 1919 richer in species than the *higher* parts. Hence it is difficult to believe that „on climbing „the mountain-slopes the number of different phanerogamous „plants becomes *gradually* smaller". All the slopes should be carefully investigated before one ventures forth such a statement.

It is curious that in this paper on galls not any mention is made of the gall-flowers of the 10 species of *Ficus* collected on Krakatao. Yet it can hardly be believed that in 1919 none of these species lodged gall-wasps in its figs. Possibly the gall-wasps bring about fructification [1]) in which case they are of importance for the problem of the revegetation. This matter ought to have been looked into. Mr. Dammerman (See p. 277) has paid some attention to gall-wasps.

B. – Dr. W. Docters van Leeuwen, The Galls of the Islands of the Krakatau group and of the Island of Sebesy in Bull. Jard. Bot. Buitenzorg, Série III, Vol. IV (1922), p. 288 seq.

It has become a time-honoured custom to begin a paper on the new flora or fauna of Krakatao with stating something like that in 1883 „the island was completely devastated, stripped of all „animal and vegetable life". That custom has also been followed here. – After having given this statement Mr. Docters van Leeuwen describes an experiment, made by him on Verlaten Eiland with one single sort of galls, with these words: „I was dwelling in a tent at the edge of the saltwater lake, „whose salinity does not differ much from that of seawater. In „the vicinity of this place the gall-midge gall on *Clerodendron* „*inerme* was very prevalent. I gathered a number of twigs „studded with galls, tied them together and then let them float „about in the lake for a week fastened by a rope to a pole „in the lake. A part of them was regularly above water, another „part was regularly submerged, and was tossed about, though „not so briskly as in a billowy sea. After this week all the galls „were stil normal, they had remained green and the gall-creatures „inside the galls were very much alive. I think I may assume „that also other species may keep alive for a shorter or longer „time in the sea, when enclosed in their galls. Ernst...... has

[1]) For one of the species occurring at present on Krakatao, *Ficus fulva Reinw.*, wrongly called by Treub *Ficus hirta Vahl*, this eminent savant has shown [Ann. Jard. Bot. Buitenzorg XVIII (1902), p. 124 seq.] that the embryo results from parthenogenetic development though germinated grains of pollen may be found on the stigma.

„already discussed the point that, wind and weather permitting, „the distance from Java and Sumatra to these islands may be „traversed within a few days by things carried by a current in „the sea. As to the distribution of other kinds of gall-creatures „we are still in many respects in the dark".

Ernst does not say that transport from the surrounding islands to Krakatao is possible within a *few* days, but within *less than a single day* [1]). I too think it quite possible that some „gall-creatures" may have been carried to Krakatao by sea-currents.

In this paper 10 galls new for Krakatao are described, found on 9 different plants. Like in the former paper not any attention is paid to the gall-wasps occurring in the figs of the many species of *Ficus* occurring on Krakatao. Yet these wasps are perhaps the only gall-insects of some importance for the revegetation of the island.

C. – Dr. K. W. Dammerman, The Fauna of Krakatau, Verlaten Eiland and Sebesy, in Treubia III (1923), p. 61 seq.

In this paper Mr. Dammerman asserts that Krakatao was in 1883 „overshed" by hot ashes forming a layer of 30–60 m. thickness and that the bottom layer of the ashes remained hot for days. Apparently he has not read the classical work of Verbeek (see p. 20) who expressly states that near the top of the cone the layer of eruptive products was very thin and who two months after the eruption found the original soil lying bare in the eroded ravines. I have already argued that the layer of ashes in the upper half of the mountain need not have been very hot. Hence it is by no means improbable that a number of small animals (earth-worms and ants f.i.) either living in the soil or taking a refuge there, has survived the eruption.

But very little is said by Mr. Dammerman on the vegetation. During his visit on October 24th 1921 all the specimens of *Carica papaya L.* „on the northern part of Verlaten Eiland „had been destroyed by a root disease. Not a single tree was „left erect, all had fallen down, bent just at the foot which was „wholly rotten". – I mention this statement because it is possible that some time or other the same disease may befall and destroy this species on Krakatao.

[1]) Ernst, Neue Flora Vulkaninsel Krakatau (1907), p. 56.

Mr. Dammerman found 3 species of fig insects on Krakatao and Verlaten Eiland. Unhappily he does not tell in the figs of which species of *Ficus* he found them, but he asserts that „all fig-trees on the island abound with fruit and in nearly „every ripe fig one will find the insect". It is certainly possible that at least some of these insects bring about fructification [1]), hence are of importance for the distribution of the many species of *Ficus*.

D. – Max Fleischer, Beitrag zur Laubmoosflora der Vulkaninsel Krakatau in Ann. Jard. Bot. Buitenzorg, XXXIII (1923), p. 105 seq.

In this short paper Mr. Max Fleischer takes the classical view-point that by the eruption of 1883 Krakatao was entirely deprived of its vegetation. This point can be passed over here, as it was already sufficiently discussed in Chapter III. Mr. Fleischer records for Krakatao 18 *Musci*, collected by Mr. Docters van Leeuwen on different excursions (1919; Jan.1922; Febr.1922). Two of these species, *Philonotis revoluta Lac.* and *Thysanomitrium Blumii Card.* were found „auf unbewachsenen Abhängen", by which perhaps steep naked ravine-sides are meant. These species may quite well have survived the eruption of 1883. It is remarkable that the 2 species found by Ernst in 1906 do not appear in the list. Assuming that they had been correctly named, they seem not to have been found back.

Most of the species were epiphytes and it is perhaps probable, though not proven, that these at least were introduced after the eruption. Several of them are quite common in Java, some seem to be rare in that island, but one should never forget that the moss-flora of many parts of Java is as yet but very imperfectly or not at all known, especially as regards the distribution of the species, many regions never having been investigated. Still less is known of the moss-flora of South-Sumatra from which as Fleischer supposes, also species may have passed over to Krakatao. – Many *Hepaticae* were found, but these seem not yet to have been named. – As to the manner of introduction of the *Musci* and *Hepaticae* found, nothing at all is known; this most interesting part of the problem has remained unsolved and probably will remain so for ever. As I

[1]) See also p. 275, footnote 2.

have already remarked before, *an oecological problem cannot be solved by indiscriminately collecting some plants.* Accurate observations and experiments by clever and well-trained botanists are needed. As regards Krakatao such observations and experiments are as yet entirely lacking.

E. — Dr. W. Docters van Leeuwen, The Vegetation of the Island of Sebesy, situated in the Sunda-Strait, near the Islands of the Krakatau-group; in the year 1921, in Ann. Jard. Bot. Buitenzorg XXXII (1923), p. 135 seq.

In this paper Mr. Docters van Leeuwen gives the botanical results of a trip made by him in April 1921 to Sebesy, an island about the size of Krakatao and lying ± $19^1/_2$ km. to the north of it. It is 43 km. distant from the nearest point of Java. Its vegetation suffered heavily from the eruption of 1883 but much less than that of Krakatao.

The results of this trip are of not the slightest importance for the solution of the problem of the revegetation of Krakatao and will therefore be passed over here. But in his paper Mr. Docters van Leeuwen gives some additional data on his trips to Krakatao. Of these data I give beneath a recapitulation paragraphed by me in order to facilitate discussion. Here again I adhere to the original English text as closely as possible.

§ 1. We are now fairly well informed as to the restoration of the flora of Krakatao. Considering the rich flora which now exists at Krakatao it has sometimes been doubted whether all vegetable life had been killed by the eruption, thus questioning the reports of the first investigators such as Verbeek and Treub and also of Van Sandick[1]). As an instance is given the occurrence of a big *Cycas*-plant which was found on the excursion described by Ernst, for it was supposed that the growth of Cycas-plants was slow. But as a proof to the contrary Mr. Docters van Leeuwen cites the occurrence of the two *Cycas*-plants discussed already on p. 217.

On this paragraph but little need be said. Of the manner of restoration of the flora of Krakatao we know *next to nothing* as was already sufficiently expounded in the preceding chapters. The papers of Verbeek and Treub were already discussed in Chapters III and IV. I was not able to consult the book of Mr. Van Sandick (who was a retired chief engineer) but judging from its title („In the „domain of Vulcanus") it is a popular work not claiming scientific authority. I do not know whether Mr. Van Sandick has ever ascended the cone of Krakatao, nor does this matter much as it

1) R. A. Van Sandick, In het Rijk van Vulkaan. Zutfen 1890 (or 1891 ?), p. 186.

is certain that he has never instituted such close and protracted investigations over the whole of the island as are required for warranting the conclusion that really the whole flora (including subterranean organs and buried seeds and spores) was annihilated. To my knowledge the presence in 1906 of a big *Cycas*-plant on the beach was never yet cited by a serious scientist as an argument against the destruction of the former flora and never will be, because we know that the beach on the south-eastern side of the island has been formed *by* or *after* the eruption, the old beach having been buried by a very thick layer of ashes and pumice.

§ 2. There remains the possibility that seeds or rhizomes of plants that grew on Krakatao before the eruption should have been preserved but this would only have created a possibility of the development of plants that can grow on bare slopes. Moreover the first pioneer plants mentioned by Treub were such whose seeds can be easily distributed by wind or sea-currents, neither does one as yet find genuine forest-plants mentioned in Penzig's article. It is difficult to assume that these forest-plants on Krakatao should have been spared as seeds or rootstocks for so many years and should have been able to sprout again when circumstances so long afterwards favoured their growth. For this proofs would have to be adduced.

In Chapter III I mentioned already that Verbeek, two months after the eruption, found the original soil lying bare in the eroded ravines. It is a well-known fact that very many seeds and rhizomes can stand being buried for a very much longer time than two months, even for many years [1]). Hence it is not at all difficult to assume that plants or seeds of the original flora may have been spared, though on this point nothing definite is known. We know also that Treub and Penzig investigated only a small part of the north-western side of the island and not the ravines on the south and south-eastern side. Hence from their finds no conclusion may be drawn as to the vegetation of these ravines. Moreover Treub saw from a long distance rather large plants growing near the top of the mountain. Nobody knows which these plants were. All these facts were already discussed in Chapter III to which I refer. I only repeat that not a single proof has been given for a sterilisation of the soil.

§ 4. The flora of Sebesy, where the destruction has not been so thorough as on Krakatao, as witness e.g. the far thinner layer of ashes that settled on this island, and where Mr. Docters van Leeuwen found a very differently composed flora (more nearly resembling what is to be found on similar

---

1) See A. J. Ewart, On the Longevity of Seeds in Proceedings Royal Society Victoria XXI (1908).

grounds on Java and Sumatra) is to him additional proof of Krakatao having indeed been absolutely destroyed.

As it appears from Mr. Docters van Leeuwen's paper he found in the localities investigated on Sebesy indeed a much larger number of species than in the examined parts of Krakatao. But as nothing is known of the flora of the two islands *before* the eruption, I do not see how the greater number of species found on Sebesy can *in any way be a proof for the* **total** *destruction of the flora of Krakatao*. At best one can say that the vegetation of Krakatao has in 1883 *suffered much more* than that of Sebesy. This is quite conceivable considering that „generally speaking the large and heavy „blocks of pumice fell inside a circle of 15 km. radius around „Krakatao. The materials fallen outside this circle consist for the „greater part of ashes. On Sebesy pieces of the size of a head are „already rare [1])".

§ 4. On Krakatao a small number of species is found to prevail in huge quantities of each. The most striking instance is *Cyrtandra sulcata* which completely and densely covers all slopes between 300 and 800 m.

For the value of this remark on *Cyrtandra sulcata* see p. 215. Mr. Docters van Leeuwen does not tell us which are the other species „prevailing in huge quantities" on Krakatao. Perhaps he means the species named in § 5, p. 207 but it is hardly to be believed that all these were very numerous. Of *Radermachera glandulosa Miq.* f.i. he himself says (See. p. 260, No. 62, herebefore) that it occurred in scattered specimens. For *Ficus variegata* Bl. see p. 284, sub h. In 1908 *Ficus fistulosa* was very common in the investigated lower ravines and *Saccharum spontaneum* formed a broad and very dense girdle on the lower parts of the south-eastern and eastern sides of the island. On p. 283 (sub e) *Macaranga tanarius* and *Pipturus incanus* are mentioned as common trees in the grassy jungle. Both species were already present there in 1908.

§ 5. After the eruption Krakatao was practically left to itself. — Then follows on the stay of Mr. Händl on Krakatao the erroneous statement already cited on p. 202, footnote 1.

§ 6. April 26th was spent by Mr. Docters van Leeuwen in an excursion to Krakatao and Verlaten Eiland.

Of this excursion nothing more is told.

[1]) Verbeek, Krakatau, Dutch Edition (1888), p. 122. See p. 23, footnote 6.

§ 7. At Krakatao Mr. Docters van Leeuwen made collections during three separate excursions; he has been three times to the top of Mount Rakata. In his earlier article on the Krakatao-flora which should be regarded as merely a provisional sketch of that flora, the harvest of the first excursions (April 24–27th 1919) has been discussed. Since then a great many new plants especially *Fungi, Musci, Hepaticae* and *Lichenes* have been noticed. At the end of Mr. Docters van Leeuwen's paper 2 lists of plants are found. The first list gives the plants found on Sebesy with mention of the kind of finding-place. In the second list have been catalogued in two parallel columns the plants of the Krakatao-islands and those of Sebesy, so as to afford a ready survey of the two floras and make a comparison possible. The Krakatao list contains all the plants thus far collected here by Mr. Docters van Leeuwen himself, so that this list also gives a complete picture of all the plants found up to date. Mr. Docters van Leeuwen avails himself of the opportunity of getting this list also printed, because he intends to watch the Krakatao-flora for a number of years before proceeding to a definite elaboration of his notes relating to the flora of Krakatao.

On this paragraph but little need be said. As appears from his publications Mr. Docters van Leeuwen was on Krakatao during 4 days in April 1919, during some hours in October 1919, during some days in April 1920, during part of a day in April 1921, during at least 5 days in January 1922 and at least one day in February 1922. We are not told whether in ascending to the top of the mountain he always followed the same path or cut different tracks. The paper on Sebesy bears no date but was probably written in 1922, 3 years after the „provisional sketch". The fact that in these few years a great many new plants were found does not in the least prove that these plants had newly settled on Krakatao but only that the exploration of 1919 was as incomplete as all former investigations and did not allow of any reliable conclusion as to the absence of a species. The lists of plants will be shortly discussed at the end of this Chapter.

§ 8. On the problem as to how the plants came to Krakatao the light has been shed that was still possible. It is now no longer possible to form anything more than an opinion on the spread of most of the plants. The three factors, sea-currents, wind and animals (including man) are known to have each of them played an important part, and we shall have to leave it at that. But the vegetation is as yet anything but normal, though several plant-associations have already been formed, but these are in most cases still very poor in the number of different species; others like the vegetation of *Cyrtandra sulcata Bl.*, which largely covers Mount Rakata from about 300 meter to the top, are very unnatural, so much that even in the three years in the course of which Mr. Docters van Leeuwen has taken stock of the island, changes have been distinctly noticeable which point to this vegetation being only of a temporary character; in the same way as the savanna vegetation of tall grasses, which clothed nearly the whole of the island at the time of Penzig's visit as it still does large areas now, has had to yield place to the mixed forest which will eventually develop into the complex old wild-forest. The main object of further investigation must be to record these alterations in the plant formations and their competetive struggle, as far as possible at regular intervals.

One is at some pains to ascertain what may be the meaning of the above statement, which is far from lucid. I pointed out already that, with the exception perhaps of the littoral flora, we do not know of a single plant how it has come to Krakatao, *if* it might have come after the eruption. We do not know with certainty whether animals and wind have played *any* part in carrying seeds to Krakatao, much less whether they have played an *important* part; one can only guess. The assertion that the vegetation on Krakatao is anything but normal, is quite enigmatical: perhaps Mr. Docters van Leeuwen confounds here *normal* with *stable*. As a matter of fact it is *quite normal* that a comparatively young secondary vegetation in the tropics is very unstable: new species constantly appearing and many old species diminishing in number or disappearing altogether. It is not clear why Mr. Docters van Leeuwen asserts that in 1922 several plant associations had already been formed. *From the very first* such associations have been developing. Treub found them already in 1886, within 3 years after the eruption. In almost every unstable vegetation associations may be observed. Even in localities where in severe east-monsoons the weeds are almost entirely destroyed by the prolonged drought, the sand-dunes of Madura [1]) for instance, they form in the next west-monsoon new and typical associations, which develop whithin a few months. The more so the plants on Krakatao, which all the year round find favorable conditions for their undisturbed development. Why the vegetation of *Cyrtandra sulcata* is called unnatural is not clear, as it has not been planted by man but was formed by natural causes. Perhaps Mr. Docters van Leeuwen means uncommon. He mentions as a curiosity that in the three years during which „he has taken stock of the island" the vegetation has noticeably changed. But every undisturbed young secondary vegetation in the tropics rapidly changes.

As Penzig examined only the beach at Zwarte Hoek and some slopes on the north-western side, it is not at all proven that at the time of his visit tall grasses „nearly clothed the whole of the „island". In the short time Penzig stayed on Krakatao he could not possibly have investigated more than a very small part of it (See p. 88).

1) See p. 72, footnote 1.

Mr. Docters van Leeuwen predicts that „the mixed forest „will eventually develop into the complex old wild-forest". If by this singular prophecy it is meant that the present forests of Krakatao will get gradually older and richer in species, I can agree with him. But I also think that there is no need for further investigations on this difficultly accessible island, Krakatao being for the botanist not more important than any secondary forest on Java which can be investigated much more easily and at much less cost.

§ 9. The single excursion to Sebesy yielded (in 6 days) 359 species whilst on the many Krakatao-excursions 259 were collected. It should be borne in mind that on Krakatao man has played a very modest part as a transmitter and a much more important one on Sebesy.

In the second appendix to Mr. Docters van Leeuwen's paper there are indeed enumerated 259 species of vascular plants for Krakatao and Verlaten Eiland together, but it does not appear which of these species were collected on Krakatao itself. Of the part man has played on Krakatao as a „transmitter" nothing definite is known, though it is probable that at least the 13 ruderal weeds found for the first time in 1919 were introduced by man or his domestic animals. Other plants also, f.i. Cocos *nucifera, Carica papaya, Citrus,* may be quite well have been introduced by human intervention.

§ 10. On the plant associations of Krakatao the following remarks may be made:

a. As to the littoral vegetation: nearly all these plants belong to the typical beach-plants; this vegetation covers a very large area and is accordingly very rich in species and in individuals.

b. the *Barringtonia* formation is confined to very narrow belts behind the first zone of plants.

c. the *Casuarina*-formation is found in several places mostly already fairly old with few rejuvenated parts. Little forests are found dotted about on top of the old ash-layer more than 80 feet above the sea; a proof of the fact that the seeds may be spread not only by the sea-currents but also by winds.

d. A Mangrove-vegetation does not exist at all.

e. the Savanna or grass-vegetation is still everywhere to be found on the hills of Krakatao but in several places and especially in the ravines it has been entirely ousted by a very young forest consisting of a very few species of trees. Especially *Macaranga tanarius M. A.* has an important effect in strangling a grass-vegetation. The seeds of this species of tree were found in the intestine of a bird, *Calornis strigatus Horsf.* (= C. chalybea Horsf.) shot on Krakatao in April 1919. These very seeds were sown by Mr. Docters van Leeuwen in the Buitenzorg garden and grew up into specimens of this tree. *Ficus* species and *Pipturus incanus Wedd.* occur numerously in these groves,

f. A forest of big heavy species of trees does not exist on Krakatao; big lianas are also missing.

g. the epiphyte flora is far richer on Krakatao than on Sebesy, not only that of higher plants such as orchids and ferns, but also of mosses.

h. *Ficus variegata* only occurs sporadically on Krakatao and of smaller dimensions than on Sebesy.

On this paragraph I can be very brief.

a. By littoral vegetation Mr. Docters van Leeuwen obviously understands only the *Pes caprae*-formation. It is to be regretted that he does not mention the species of which it is „so very rich." Now his assertion is without value. For his list of littoral plants given in the Transactions of the Congress (1920) I refer to p. 234.

b. In 1908 the *Barringtonia*-formation on Krakatao was frequently lying immediately behind the floodmark (and not behind the first zone of plants), wherever the beach was being washed away by the waves. There is no reason to suppose that in 1919 and later years it should have been otherwise.

c. We are not told how far the „little forests dotted on top of „the old ash-layer" were distant from the sea. It is a long-known fact needing no new proof that the light winged seeds of *Casuarina equisetifolia* and other tall trees may be carried over some distance by wind. It was f.i. already mentioned by Kamerling[1]) in 1911. See also p. 91, No. 6.

d. Krakatao possessing neither muddy shores nor coral-terraces running dry or almost so during the ebb-tide, there is no fit substratum for mangrove-plants to develop upon.

e. For the „very young" forest in the ravines, see Chapter IX in which the excursion of 1908 is discussed. From the fact that seeds of *Macaranga tanarius* taken from the intestines of a *shot* bird had retained their germinative power it may certainly not be deduced that the seeds leave the digestive canal of *living* birds likewise in an undamaged state. Even if this might be the case one may only conclude that the birds can take an active part in spreading the tree *over the island* but not that they have introcuced the plant *from elsewhere.*

---

[1]) Dr. Z. Kamerling, Over het voorkomen van Wortelknolletjes bij *Casuarina equisetifolia* in Nat. Tijdschr. Ned.-Indië LXXI (1911), p. 20.

g. A comparison between the epiphytic flora of Krakatao and Sebesy can only be made after both islands have been thoroughly investigated by a competent botanist, recognizing the plants in loco, and having a keen eye to local circumstances.

§ 11. Plants with strong rhizomes such as many *Zingiberaceae*, *Musaceae* and *Araceae* are not at all or but scantily represented on Krakatao. Mr. Docters van Leeuwen found on Krakatao one species of *Alocasia* and two *Zingiberaceae*, one of which, *Costus sericeus Bl.*, is perhaps spread by birds and the other, *Gastrochilus panduratum Ridl.* occurred in the garden of Mr. Händl who spent a short time on Krakatao.

It is true that but few species of *Zingiberaceae* and *Araceae* have been found on Krakatao. But some other terrestrial plants with strong rhizomes occur there, as *Blechnum orientale, Diplazium polypodioides* and *Angiopteris evecta.* Nothing is known about the manner in which the arillate seeds of *Costus sericeus* are dispersed [1]. This may sometimes be done by animals but it is not probable that *birds* take an important part in the dispersal, as the seeds are rather soft. *Gastrochilus panduratum,* a common plant in many teak-forests, is very frequently cultivated in Javanese gardens for its aromatic roots; the specimen found in the garden of Mr. Händl was doubtless a planted one and of no importance at all for the study of the development of the new flora, as this plant cannot thrive in the grass-jungles and rain-forests of Krakatao. The stay of Mr. Händl on Krakatao was discussed in Chapter X.

Appendix I to the paper of Mr. Docters van Leeuwen gives of the vascular plants found on Sebesy a list, which is of no importance at all for the study of the renewal of the vegetation of Krakatao. Appendix II records in the first column the plants of Krakatao and Verlaten Eiland, in the second those of Sebesy. The list of plants found on Krakatao and Verlaten Eiland does not mention on which of the two islands the different species were found. It is a mere list of Latin names, nothing more. Nothing is told about habitat, nothing about frequency, nothing about the reproductive organs. About the mode of distribution not a single observation is recorded. In short this list might have been composed by any person who had collected some plants on the islands and had them named by others. For the study of the revegetation-problem it is valueless.

[1]) Cf p. 190, No. 12.

Let me conclude this paper by comparing the litterature on the revegetation of Krakatao to „the colossal skeleton brontosaur" some American author speaks of, „that stands fifty-seven feet long and „sixteen feet high, the awe and admiration of all the world, the „stateliest skeleton that exists on the planet. There was one small „bone and the rest was built out of plaster of Paris. If the builders „had not run short of this stuff the monster would have become „much bigger".

I have tried to find the very few bones in these wagon-loads of rubbish and will finish by stating once more, that:

1. *It is not at all proven that by the eruption of 1883 all vegetable life on Krakatao was destroyed.*

2. *Even if this could be proven, we know – with the exception of the littoral flora –* **nothing at all** *about the manner in which the new vegetation has appeared. Only guesses without scientific value have been made but no reliable observations nor experiments.*

3. *Therefore the Krakatao-problem can neither now nor in the future either be posed or solved and is of no importance at all for Botanical Science.*

Pasuruan (Java), April 1929.

INDEX OF PLANT-NAMES.

Bold type: plants found on Krakatao.
Ordinary type: other plants.
CAPITALS: Names of plant-families and orders.
Italics: Synonyms.

GPSR Compliance

The European Union's (EU) General Product Safety Regulation (GPSR) is a set of rules that requires consumer products to be safe and our obligations to ensure this.

If you have any concerns about our products, you can contact us on ProductSafety@springernature.com

In case Publisher is established outside the EU, the EU authorized representative is:

Springer Nature Customer Service Center GmbH
Europaplatz 3
69115 Heidelberg, Germany

Batch number: 10824743

Printed by Printforce, the Netherlands